Writing Projects for Mathematics Courses

Crushed Clowns, Cars, and Coffee to Go

© 2004 by

The Mathematical Association of America (Incorporated)

Library of Congress Catalog Card Number 2003113542

ISBN 0-88385-735-9

Printed in the United States of America

Current Printing (last digit):
10 9 8 7 6 5 4 3 2 1

Writing Projects for Mathematics Courses

Crushed Clowns, Cars, and Coffee to Go

Annalisa Crannell
Franklin & Marshall College

Gavin LaRose
University of Michigan

Thomas Ratliff
Wheaton College

Elyn Rykken
Muhlenberg College

Published and Distributed by
THE MATHEMATICAL ASSOCIATION OF AMERICA

CLASSROOM RESOURCE MATERIALS

Classroom Resource Materials is intended to provide supplementary classroom material for students—laboratory exercises, projects, historical information, textbooks with unusual approaches for presenting mathematical ideas, career information, etc.

101 Careers in Mathematics, 2nd edition edited by Andrew Sterrett
Archimedes: What Did He Do Besides Cry Eureka?, Sherman Stein
Calculus Mysteries and Thrillers, R. Grant Woods
Combinatorics: A Problem Oriented Approach, Daniel A. Marcus
Conjecture and Proof, Miklós Laczkovich
A Course in Mathematical Modeling, Douglas Mooney and Randall Swift
Cryptological Mathematics, Robert Edward Lewand
Elementary Mathematical Models, Dan Kalman
Environmental Mathematics in the Classroom, edited by B. A. Fusaro and P. C. Kenschaft
Essentials of Mathematics: Introduction to Theory, Proof, and the Professional Culture, Margie Hale
Exploratory Examples for Real Analysis, Joanne E. Snow and Kirk E. Weller
Geometry from Africa: Mathematical and Educational Explorations, Paulus Gerdes
Identification Numbers and Check Digit Schemes, Joseph Kirtland
Interdisciplinary Lively Application Projects, edited by Chris Arney
Inverse Problems: Activities for Undergraduates, C. W. Groetsch
Laboratory Experiences in Group Theory, Ellen Maycock Parker
Learn from the Masters, Frank Swetz, John Fauvel, Otto Bekken, Bengt Johansson, and Victor Katz
Mathematical Evolutions, edited by Abe Shenitzer and John Stillwell
Mathematical Modeling in the Environment, Charles Hadlock
Mathematics for Business Decisions, Part 1: Probability and Simulation (electronic textbook), Richard B. Thompson and Christopher G. Lamoureux
Mathematics for Business Decisions, Part 2: Calculus and Optimization (electronic textbook), Richard B. Thompson and Christopher G. Lamoureux
Ordinary Differential Equations: A Brief Eclectic Tour, David A. Sánchez
Oval Track and Other Permutation Puzzles, John O. Kiltinen
A Primer of Abstract Mathematics, Robert B. Ash
Proofs Without Words, Roger B. Nelsen
Proofs Without Words II, Roger B. Nelsen
A Radical Approach to Real Analysis, David M. Bressoud
She Does Math!, edited by Marla Parker
Solve This: Math Activities for Students and Clubs, James S. Tanton
Student Manual for Mathematics for Business Decisions Part 1: Probability and Simulation, David Williamson, Marilou Mendel, Julie Tarr, and Deborah Yoklic

MAA Service Center
P. O. Box 91112
Washington, DC 20090-1112
1-800-331-1622   fax: 1-301-206-9789

Contents

1

Introduction and Supporting Information

1.1 Introduction

In this volume we have collected a group of writing projects suitable for use in a wide range of undergraduate mathematics courses, from a survey of mathematics to differential equations. The projects vary in their level of difficulty and in the mathematics that they require, but are similar in their mode of presentation and use of applications. Students see these problems as "real" in a way that textbook problems are not, even though many of the characters in the projects (e.g., dime-store detectives and CEOs) are obviously fictional. The stories that these characters tell are sometimes fanciful and sometimes grounded in standard scientific applications, but the mere existence of the story draws the students in and makes the problem relevant.

This volume is the combined effort of four instructors who have used writing projects in their own courses for six or more years. Our goal is to provide an easy-to-use, widely applicable set of course materials that instructors can adopt and adapt to their courses—as many others who have come upon the projects through contact with the authors or the authors' Web sites have already done. Toward that goal, we include not only an extensive set of class-tested projects, but also

- implementation notes which highlight student difficulties,

- information about project solutions, and

- advice for grading writing projects.

In the remainder of this chapter we discuss how we have successfully used and graded the projects. In Chapter 3 we include sample solution papers for two of the projects. This collection, taken in toto, addresses the traditional difficulties encountered when using writing assignments—those of coming up with good projects, implementing them effectively, and grading them efficiently.

1.2 Motivation

Why should students write papers in math classes? In particular, why should students write the type of papers we present in this book?

We want our students to perceive mathematics as useful in their academic careers and in their lives after college. We want them to see writing in the same light. In more traditional mathematical courses, however, students are often left with simple exercises that are devoid of context. These exercises do little to convince students that they will be able to use interesting mathematics in their own lives, or that writing about mathematics has any purpose beyond the classroom. In contrast, the projects in this volume abound with plot and character. They provide genuine applications that our students have commented on with appreciation.

Plot and character help students to focus their thoughts by presenting them with a reason for responding with a clear written solution. When a student writes a lab report or a homework assignment, the student explains mathematics to somebody who already understands all the material very well (the instructor). Worse yet, the student knows that the instructor can solve the same problem more quickly and easily than the student can. In this situation, writing becomes a question of putting down enough information to get partial credit. On the other hand, the projects in this book are cast in the form of letters from well-defined (if fictional) characters who have no idea how to solve the problem. (Indeed, sometimes the character is not entirely sure how to *state* the problem.) Students who write in response to a character therefore have a well-defined audience, which allows them to establish the level of mathematical sophistication and written explanation that fits the needs of this character.

We also firmly believe that the act of writing mathematics can itself enhance student learning. The depth of understanding required to produce a lucid mathematical explanation is generally deeper than that demanded by traditional homework assignments. As our students write their solutions, and therefore as they discover how much harder it is to explain than merely to solve the problem, they deepen their own understanding of the mathematics underlying their solution. This increased understanding is independent of whether the same material had been covered in class or homework assignments.

Finally, projects are by their very nature open-ended and therefore difficult, demanding that students develop and practice the problem-solving skills that are the hallmark of mathematics. These skills are the reason that many students dread mathematics, and at the same time are the reason that many students are required to take mathematics outside of their majors. Perhaps the most effective way of developing students' ability to analyze and solve difficult problems is to challenge them to do so, and these projects fulfill that role. In addition, because these problems take days or weeks instead of minutes or hours to solve, students are taught not to give up easily, but to develop patience.

Given these benefits to the students, the obvious drawback to the instructor is the amount of time that is required for grading and commenting on written assignments. A colleague of one of the authors was once heard to say, "I have yet to see a teaching technique improving student learning that took less time from the instructor." The time and effort demanded by the addition of student writing to a course is certainly at first blush a daunting prospect. However, we assert that with the supporting materials in this book, the sweat and tears are less than they appear at first sight, and the benefits are manifold.

1.3 Use and Implementation of Projects

There is no one way in which these projects should be used in a class—because of the personal nature of successful teaching, the method and course organization that works for one instructor may be very different from that successfully used by another. We include the following "models" to show some of the ways in which we have used the projects, illustrating at least some of the different, successful manners in which they may be used.

These models, while different, do share some common threads. In each case, the student's solution to the project is a written paper that describes in precise mathematical prose how the problem was resolved. The students are expected to do their own work (either in their group or by themselves), asking questions of the instructor (or possibly other students) when they get stuck—in no case is it expected that a submitted solution paper will demonstrate a complete lack of understanding of the problem or the mathematics behind its solution. But there are also differences. Some instructors have the students work in groups to solve the problems, others may have them work with other students but have each student hand in an individual solution to be graded, and some have them work individually.

1.3.1 Annalisa Crannell's Model

Annalisa Crannell teaches at Franklin & Marshall College, a highly-selective liberal arts college of 1800 students. She uses writing projects in all three semesters of calculus. Her students are mainly freshmen; they have not declared a major but have a very slight inclination toward the sciences; the class size is usually 25 or fewer.

Annalisa introduces these writing assignments by declaring (with a straight face) that this is how mathematics is "done" in college. She assigns three projects a semester. The first project must be written up individually and counts for 5% of the final grade; students may choose to write the second and third projects by themselves or with a partner of their choice, and each of these projects counts for 10% of the final grade. Students have two to three weeks to work on each project; they may see the instructor outside of class for help, but there is little or no class time devoted to the projects.

Annalisa requires her students to use the computer to write their solutions, and she grades their work with the use of a checklist (see Chapter 3). In fact, in an effort to focus on the process of writing she returns only the checklist and not the student's project—although the student has a copy on the computer, presumably.

She grades all three projects using the same criteria (using the same checklist), so that the average grade on the first project is usually between a 'D' and a 'C', but the final project is considerably improved—usually in the high 'B' or low 'A' range.

1.3.2 Gavin LaRose's Model

Gavin LaRose used this model at Nebraska Wesleyan University, a small (1500 students) moderately selective liberal arts university, in calculus, differential equations, and linear algebra classes ranging in size from 12 to 25. He assigns two or three projects in a semester. The projects are worked on by groups of two or (if there is an odd number of students in the class) three students. Because the groups produce a single solution paper, the choice of group size seeks to minimize the likelihood that one student in the group will "coast," allowing his or her partners to do most of the work. Gavin assigns students to the groups with the intent of (1) never having the same students work on more than one project together, and (2) pairing students with similar work ethic or ability.

Gavin hands out the projects two or three weeks in advance of the due date, and requires that groups meet at least briefly (for a couple of minutes) with him approximately a week after they have received the project to ensure that they have begun work (as an extension to this, they may also be forbidden from asking any questions about the project over the weekend before it is due). The projects each constitute between five and seven percent of the students' course grade, so that three projects will determine 15–20% of their grade. Because the projects are group assignments, with all members in the group getting the same grade, this has the effect of raising all grades somewhat and reducing the spread of the grade distribution. Other homework in the course is reduced slightly when the most fevered work is being done on the projects, but no formal class time is dedicated to the projects—though in lab classes a lab period may be declared an open time slot in which students are encouraged to work on (and ask questions about) the project.

The actual solution paper is required to be "long enough," which usually means 3–7 pages depending on the course and the project. Gavin requires that they be word-processed (though in lower-division courses hand-written equations are acceptable) and in the format of a technical paper. Instructions on how to write in this manner are given verbally, and he has sample papers available for students to examine. Student papers are graded using a rubric, and the scores on the rubric returned to the students with their marked paper.

1.3.3 Tommy Ratliff's Model

Tommy Ratliff teaches at Wheaton College, a selective liberal arts college of 1500 students in Norton, Massachusetts. He uses writing projects in the three semester calculus sequence and Linear Algebra. He assigns three projects per term in Calculus I and II and two per term in Multivariable Calculus and Linear Algebra. The class size is approximately 30 students, and the projects typically count for 20% of the final grade. The students self-select into groups of two or three for the semester, and each group turns in a single paper for each project. The students have two class meetings to work on each project in the courses that meet four times per week (Calculus I and II) and one class meeting per project in the other courses. The papers are due approximately one week later and are typically 5–10 pages in length. Other than the class time devoted to the projects, there is no adjustment in workload made for the projects. Many of Tommy's projects are not completely self-contained and require extra help from the professor. This is by design. He wants the students to wrestle with the problems, and if there is a hint that is useful for several groups, he will take a few minutes to speak to the entire class during the class meetings when they are working on the projects.

Tommy also uses a checklist to grade the projects and allows each group to determine how to distribute the project grade among the group members. For example, if a group of three receives an 80 on their paper, then the group determines how $3 \times 80 = 240$ points should be distributed among the members. This has worked surprisingly well, and he has had to mediate this process only twice in four years.

1.3.4 Elyn Rykken's Model

Elyn Rykken currently teaches at Muhlenberg College, a highly-selective liberal arts college of 2000 students. She uses writing projects in her calculus courses. Her students are mainly freshmen and sophomores who have declared science majors; the class size is 25 or fewer. Before coming to Muhlenberg, Elyn taught at a mid-size (5000), minimally selective state university. There she used writing assignments in a basic mathematics course. The course was a terminal mathematics course designed for non-science majors. The prerequisite for the course was minimal competency in algebra.

At Muhlenberg, Elyn assigns two or three projects a semester. The students are allowed to work in groups of up to three people and allowed to hand in a joint paper. At her previous institution, Elyn assigned three or four projects a semester. The students worked to produce independent solutions (possibly with some collaboration with other students). Each student was responsible for his or her own write-up.

The projects are always handed out two to three weeks before they are due, and each counts for roughly 4%–5% of the total course grade. Elyn also uses a checklist to grade the projects. During semesters where four projects are required, the lowest score is sometimes dropped so that total percentage of the final grade is 15% or less. No formal class time is dedicated to the projects and students are encouraged to work together and ask the instructor for help outside of class.

While there is no set length required for the papers, they usually range from two to five pages. Typing is not required, but it is strongly encouraged. At Muhlenberg, the grades usually fall in the A to low C range. The grades were lower at her previous institution. Some otherwise strong students do earn lower marks for their projects, but overall, the projects do not seem to significantly change a student's grade.

1.4 Grading Project Solutions

The task of grading the solution papers for the projects is, perhaps, the largest impediment to their use. However, we have found that this is expedited considerably through the use of rubrics or checklists, which are explained separately in the following sections. While these two methods are similar in their general intent of establishing a set of criteria that may be evaluated with relative ease from a read of the papers, they differ in implementation specifics, as we describe below.

1.4.1 Checklists

Annalisa, Elyn, and Tommy all use checklists to grade the projects. We find that the checklists help us to grade consistently from student to student and also from assignment to assignment. In addition, the checklists help us minimize grading time. (It is much easier to write "no" next to "Does this paper clearly (re)state the problem to be solved?" than to write out the phrase "You must clearly restate the problem to be solved.")

Our students like having these checklists. These sheets of paper not only provide the students with useful feedback after their papers are returned, but also guide the students before they submit their work. (Students are more likely to clearly explain the problem to be solved if their checklist says they must.) The checklists serve as a grading tool, but they also serve to teach students the basic techniques of writing a scientific or mathematical paper.

Annalisa's checklist is given in Chapter 3 with a sample student paper. Tommy's and Elyn's checklists, which are in Chapter 4, differ slightly; for example, Elyn assigns different point values to different items in her checklist. Each of us began by making a list of criteria we look for in good technical writing; we've tweaked our lists over the years as we read student work. We hand out these checklists along with the writing assignment; students may—in fact, *should*—use the checklists as they write, and submit the checklist along with their paper. We use the checklist to grade the paper and to provide feedback. Then we return the checklist (with or without the student's paper) to the student.

The use of checklists in grading student papers is described in greater detail in the article "How to grade 300 mathematical essays and survive to tell the tale" [A. Crannell, *PRIMUS* 4(3):1994].

1.4.2 Rubrics

Gavin uses rubrics to grade projects. The rubric seeks to isolate the different required components in the solution to the project, and to assign to each of these some number of points (usually two to four) to which the student(s) is(are) entitled if that component is present and clearly explained in the solution paper. As an example, consider the project "A Jump and a Jerk," for which students must derive, based on information provided, an equation modeling the motion of a falling body. The modeling equation is $F = ma = mv'(t) = mg - Kv(t)^2$, and for the derivation of this the rubric might give 0 to 3 points as shown in Table 1. The complete rubric is in Chapter 4.

TABLE 1
Sample Rubric Objective

Objective	0 points	1 point	2 points	3 points
Model development	no sensible model	one term (mv' or a force) in model correct	two terms correct, or three but with errors in finding parameters	correct model, well explained

For a typical project there are generally on the order of four or five steps (similar to that illustrated in the sample rubric objective in Table 1), so that by assigning each of these up to three or four points and giving some credit for a clear paper and meeting all appropriate deadlines the project is scored out of 15 or 20 points. Once the rubric is determined, grading a project using it usually involves reading through the project once and then going back to spot check the success with which students have met the requirements for each of the steps in each objective included in the rubric. A complete rubric appears in Chapter 3 with the sample graded student solutions.

The creation of the rubric itself, when thought out in the level of detail presented here, clearly represents an initial outlay of time before any given project can be graded. However, this pays off in spades when grading the papers, as it is then possible to grade any given paper with one reasonably quick read through it. Finally, it is worth noting that for some projects it may not be necessary to articulate in such detail the requirements for each of the categories given in Table 1.

Unlike the checklists, it is clearly impractical to give the rubric to the students with the project, because it details the solution that they are to arrive at independently. However, it is easy to summarize the rubric used when the graded papers are returned to the students.

2

Projects

In this chapter, we present three dozen writing projects that we have used in mathematics courses ranging from precalculus to differential equations. Because we (the four authors) have such different writing styles, we group the project first by author and then by the course in which we assigned the project.

Readers should note, however, that many of the projects may be applicable to other courses. For example, some of the projects from early in Calculus I would work well in a Precalculus course, and some of the Calculus I projects may be just as appropriate for Calculus II, or vice versa, depending on the syllabus of the specific course.

For this reason, we have included introductory remarks with each project that we hope will help you determine if a project is right for your class. These include

- the key concepts of the project,

- the realism of the project, if any,

- the students' reactions to the project, including parts that they found especially challenging,

- any credits for the project, and

- the minimal technology required to complete the project.

We also include a brief solution to each project. (In Chapter 3, we present two full student solutions, along with comments on grading and giving feedback.)

The Case of the Dough for the Greenhouse

Annalisa Crannell

Calculus I

Concepts　Translating from words into formulas; reading graphs

Realism　Completely fabricated

Reactions　This is the first in a six-part series. Students had a little difficulty with the square term (the "cost per tree per tree"), but otherwise found this a good warm-up writing assignment. This project may also be appropriate for a precalculus course.

Technology　A graphing calculator is helpful, but not necessary.

Growing Mature Arboreal Trees
Oak Grove Avenue
Pinneapolis, PU 22222
September 5, 1997

Calculus I Students
Franklin & Marshall College
Lancaster, PA 17604-3003

Dear Calculus Students:

I am desperate for help, and am in agony because I think no one but you can save me from this horrible mess. Indeed, it may be that even you cannot save me—that I am entirely done for—but I'm pleading with you in spite of it all. Please, please help me. I don't know where else to turn.

The trouble all began—although I didn't know it then—about five years ago when my company hired Jack Phaze. Jack was a real toady, always sucking up to the company owner, Mabel Sabbling, and sneering at the rest of us. It wasn't long before most of us called him "Jerk Face" behind his back—you never met a guy less willing to do an honest day's work, or more willing to take credit for work done by others.

It wasn't too long either before Jack surprised us by getting promoted into a position of importance. I guess I should back up and explain what our company does. The "Growing Mature Arboreal Trees" Company (or as it is popularly known, "GMA Tree") cultivates exotic trees for use in bird sanctuaries and movie sets. As you can imagine, there's very little demand for this kind of thing, so we have a small business with few competitors. If I lose this job where I've worked all my life, there's nowhere else I can go. I know exotic trees—that's all!

About a year ago, Jack finally was promoted to working directly beneath me, to my great chagrin. The morale of all the people in my group went down, absenteeism went up, and I'm sure it was all because of how miserable it is to work with Jack. One day I actually pulled aside a group of my employees to ask them what was bothering them, and they confirmed my suspicions: Jack never did what he was supposed to, and everyone else ended up covering for him just to get the projects done. Then, after they'd covered his rear end, he'd sneer at them for how hard they worked. I was very angry, as you could imagine, and I told them, "If I have anything to do with it, Jerk Face is out of here."

Wouldn't you know it, right at that moment he walked around the corner—I could tell he'd heard me by the look on his face. We've been out for each other's blood ever since.

I had a harder time getting rid of him than you might imagine. It turns out that Phaze has an uncle in the bird sanctuary business (clearly the secret of his longevity in our company).

Well, shortly after the "Jerk Face" incident, I had a visit from a herbicide specialist that Mabel Sabbling (the company owner) wanted me to talk with. Edgar (I don't remember his last name, but I wish I did!) convinced me that we could build a greenhouse which protected trees from insects and disease. He showed me a bunch of figures, enough to convince me to commit our company to eventually purchase such a greenhouse. Then, just two days after I signed the contract, he told me he'd been hired by a bird sanctuary and was being sent to Costa Rica. I haven't been able to get his forwarding address from his new company, so I can no longer ask for his help. I smell Jack behind all this!

Here's the problem. Our company can only commit $4225 to this project without going into debt. Also, since the greenhouse can only hold so many trees, we need to make sure that we're spending no more that $100 per tree—otherwise we'll operate at a loss. Ed assured me this was possible—he even showed me how to do it—but he took most of his notes with him, and I'm no mathematical genius.

Here's how the expenses add up. First, to build a greenhouse, you need $2,222 just in start-up costs. Then, for each tree in the green house, you need $5 for a proper planter. But in addition, because the more trees there are, the easier it is to spread infection, the cost of disinfecting any one tree is $1 for each tree in the greenhouse.

The one example Edgar did that I managed to save goes like this: suppose we plant 10 trees. Then we'd spend $2222 for the greenhouse, plus $50 for planters, plus $10 to disinfect each of 10 trees (meaning $100 for disinfecting the whole greenhouse). The total cost would be $2,372. So far, so good. But unfortunately this comes to over $237 per tree—that's bad.

I tried figuring out what happens if we increase the number of trees, say to 100. In that case, I got the cost per tree to be a better (but still not acceptable) $127, but the total cost to be an exorbitant $12,722. I'm not even sure these figures are right, however. I'm sure you could tell me.

I showed these figures to Sabbling, and she threw a fit—swore I'm going to wreck the company. She gave me three weeks to come up with a way to make this work, or I will be out of a job. I just know Jack is licking his lips over all this, waiting to take my place.

Can you help me figure out what to do? I'm sure that Ed was sincere, just as I am sure that Jack is a duplicitous, two-faced, no good scoundrel. I have until September 22 to get on track with Sabbling (that's when the contract I signed says we have to place our order with the company).

Yours sincerely,
George Bush
Associate Vice President for DisInfection

Solution

The total cost of the trees is $C = 2222 + 5T + 1T^2$, where T stands for the number of trees. The cost per tree is $CPT = C/T$. Graph these on a graphing calculator (choosing suitable ranges for the window), and you discover that 42 trees is just the right size; it satisfies both the total cost and the cost per tree constraints.

The Case of the New Leaf

Annalisa Crannell

Calculus I

Concepts   Riemann Sums

Realism   Real-world application with fabricated storyline

Reactions   This is the second part of a six-part series. Students began this project very nervously: how do they find an equation for the curve of a leaf? But once they discovered that they could "just" trace and count, they became artistic. The leaves that I used were close enough in size that it took very accurate measurements to decide which had the larger surface area. Although I thought that the situation was highly fabricated when I wrote this, I discovered subsequently that our biology department has a $2,000 machine which does exactly what I asked the students to do—indeed, several of the students got this project right after their leaf-measuring biology lab!

Credits   This was inspired by a project to measure the area of students' hands that appeared in the MAA volume *Student Research Projects in Calculus*, Cohen *et al*.

Technology   Photo copier

Growing Mature Arboreal Trees
Oak Grove Avenue
Pinneapolis, PU 22222
September 24, 1997

Calculus I Students
Franklin & Marshall College
Lancaster, PA 17604-3003

Dear Calculus Student,

I can't thank you enough for your help on Monday. You really saved my skin. Better yet, you managed to do me a favor that neither I nor my companions will ever forget!

You see, Monday morning when I took your figures into Sabbling's office, Phaze got there first. He had just about convinced her that the project was impossible and she should fire me without delay. Just as I walked in, he was telling her, "Ms. Sabbling, I'm so sure that this is all a scam that I'll stake my job on it. If mousy old George can pull it off, I'll resign." Well, mousy old George pulled it off with your help, and Sabbling wasted no time at reminding him of his vow. I guess his snow job didn't work as well on her as we all thought it did. What a party we had to celebrate! Thank you again!

It's probably not quite polite to ask you for help again in a thank you note, but you were so helpful last time I'm going to risk it.

Sabbling was so impressed with my (meaning your) mathematical prowess that she put me in charge of the new leaf-measuring project. There are all sorts of things we do where knowing the surface area of a leaf would help us—knowing how much insecticide to order, figuring out how much sunlight a leaf could absorb, etc. Unfortunately, this figuring out business isn't as easy as it seems.

Just as a "for example," I photocopied two leaves on the back page. One is from a White Poplar, a tree introduced during colonial times and very hardy in cities. The other is from a Spotted Oak, sometimes called a Shumard Oak after the state geologist of Texas in the mid-1800s. One of my employees, Frank, says the Oak leaf is larger—it's clearly longer. Betty says that the Poplar leaf is larger, because it's wider. I have no way of knowing who is right or by how much.

Do you know of any way of figuring out the surface area of a leaf to, say, within one square inch? If you could tell me the area of each of these leaves, I'd be most grateful. Better yet, if you could let me know how to do it for general leaves, I could have Frank and Betty do it in the future, so I won't have to bug you any more. They're both embarking on a big measuring project in 2 weeks, so getting an answer by October 6 would be greatly appreciated.

Yours most gratefully,
George Bush
Associate Vice President for DisInfection

Solution

One way to solve this is to trace the leaves onto graphing paper. Count the squares lying completely inside the leaf: this gives a lower bound on the area. Then count the squares which contain a piece of the leaf: this gives an upper bound on the area. If the squares on the paper are fine enough, then the estimates should be within a square inch of each other.

The Case of the Dropped Baton

Annalisa Crannell

Calculus I

Concepts   Constant velocity versus constant acceleration

Realism   Real-world application with fabricated storyline

Reactions   This is the third in a six-part series. I thought that this was a hard project, but the students seemed to relate well to the subject matter.

Credits   Thanks to Pete Carrol, the F&M track coach!

Technology   None required

Growing Mature Arboreal Trees
Oak Grove Avenue
Pinneapolis, PU 22222
October 31, 1997

Calculus I Students
Franklin & Marshall College
Lancaster, PA 17604-3003

Dear Calculus Students:

You don't know me—my name is Betty Kant, but you do know my boss, George. He's such a sweet guy and a wonderful boss. I can't imagine anything I'd rather do for a living than measure leaves for him. He tells me that it was you who came up with the new measuring technique—I have to tell you that that's made our job a whole lot easier.

But that's not actually why I'm writing you. I was hoping—since you're so good at solving our problems—that you could help me with a a problem that's been plaguing my track team.

We're all amateurs, and we all suffer from having no depth perception. Normally this is no problem. When you run events on the track, you just go where the track goes and stop when you cross the line—no big problem. In fact, we can all do the quarter-mile (well, 400 meters) in just about 66 seconds. Pretty respectable for a bunch of leaf measurers! The sticky part comes when we try to do relays. We're horrible at hand-offs!

The thing is, in a track relay, each runner has to hand a baton to the next runner in a short box called the passing zone, which is about 20 meters long. Let's suppose our first runner, Amanda, wants to hand the baton to me. She comes chugging down the home stretch. If I stand at the beginning of the box and wait for her, she blows by me and it takes me a while to get going. I'm allowed to start running before she gets to the passing zone, but sometimes (no depth perception) I start too soon, and then I nearly run out of the end of the box before she gets there, so I have to slow down and wait again. The same dilemma happens when I pass to Carla or when Carla passes the baton to Darlene. It's awfully embarrassing to have such lousy handoffs when we can run so respectably all alone. Darlene flat-out refuses to join the relay any more after one particularly sloppy race a year and a half ago.

But suppose we put a yellow bandana down on the ground at just the right distance before the passing box. Then when Amanda runs past the bandana, she could yell "go!" and I could run as hard as I could—I know I can get up to full speed in just 2 seconds if I don't have to worry about Amanda meeting up with me—and we'd have a near perfect hand-off. Then I could do the same for Carla, and Carla for Darlene.

Do you have any way of figuring out where we put the bandana? There's a big track meet coming up on November 17. If I could get your answer and a way to persuade doubting Darlene that it ought to work, I'd really appreciate it!

Sincerely,
Betty Kant

Solution

At the time of the hand-off, we want the two runners to be running the same speed ($400/66$ meters/second) at the same place (the same distance past the beginning of the passing zone). My students chose to define distance from the beginning of the passing zone. We now need to solve for B, the distance of the bandana (in meters) before the start of the passing zone.

We make a crucial pair of assumptions that Amanda runs at a constant speed and that Betty has a constant acceleration. This may or may not be valid: perhaps Amanda wears down by the end of her run, or perhaps she has an amazing kick at the end. But given these assumptions, we conclude that the two runners go the same speed after $T = 2$ seconds. By the time of the hand-off, Amanda has run $(400/66) \times 2 - B$ meters from the starting zone. The second runner, Betty, has an acceleration of $(400/66)/2$, a velocity of $(400/132)T$ meters/second and a distance which is $(400/264)T^2$ from the passing zone. Setting the two distances equal when $T = 2$, we see that the bandana should be placed $400/66$ meters before the passing zone.

Some people guessed that $B = -400/66$ from the beginning because of symmetry arguments. If the time it takes to get up to full speed is $T = s$ seconds instead of 2 seconds, this changes the bandana location to $B = -400s/132$ meters, a slightly less intuitive answer.

The Case of the Falling Grapefruit

Annalisa Crannell

Calculus II

Concepts   Acceleration with air resistance; asymptotes; exponents; elementary differential equations

Realism   Real-world application with fabricated data

Reactions   This was the fourth in a six-part series, but it was the first project for most of the students in the class. They found it *very* difficult, and so I gave them hints along the way.

Credits   Thanks to Hall Crannell for his rule-of-thumb estimate of the terminal velocity of grapefruits.

Technology   A graphing calculator is helpful.

The Citrus Clinic
PO Box 2358
Big City, PU 11235
January 14, 1998

Calculus II Students
Franklin & Marshall College
Lancaster, PA 17604-3003

Dear Calculus Students:

Grapefruits can only fall so fast. It's sad, but it's true.

When I was in calculus, they taught me that the speed of a falling object was just gravity times time, but they lied. I know, because I've dropped a lot of grapefruit in my day, and they've never gone quite as fast as my teachers promised me they would. Next time, I'll use a helicopter instead of a tall building, and try to get up to speed from higher up.

I suppose I should back up a bit. You probably want to know why I'd want to drop grapefruit anyway. I mean, I guess that's a normal question for someone to ask, given that most people don't go around dropping grapefruits for a living. Dave Barry might be an exception I guess, but he's more of a watermelon or Volkswagen kind of a guy. Grapefruits are less spectacular, but they're easier to carry around without arousing suspicion. Plus, they're cheaper than typewriters. They're biodegradable, too.

I started dropping grapefruit as a way of relieving stress during college. You know the feeling: midterms come around, everyone is wired on caffeine, your roommate has that psycho-killer look in his eye. I got the idea that smashing things would be cathartic. Just take something and crunch it to pieces. I discovered early on that my roommate's ceramics project was a bad idea (psycho-killer and all that). Fruit seemed to be less controversial.

The weight room at the gym was a wonderful place to get rid of all those pent up tensions, at least until the trainer caught me smashing bananas in the Universal weight machine. He didn't seem to appreciate the aesthetics of the situation, if you know what I mean. The roof of the chemistry building seemed to be a much better locale for letting my creativity run wild: in the true experimental spirit, I released various fruits. Kumquats smashed well; cherries and plums were hard to see by the time they splattered; apples and lemons held together just a little too much to be satisfactory. Twinkies just bounce. But an old, warm grapefruit did the trick. What a great smash it made! The feeling I got was all I needed to drive away those midterm demons. You had to be there.

To make a long story short, others got involved. I started a stress clinic with grapefruit therapy. It's been more successful than you'd think, really. I've started advertising with glossy brochures, and that's where I get into trouble.

I have an office in a very tall building. My clients drop their grapefruits from there, and since it takes the grapefruits 3.3 seconds to fall, I claim that they end up going "more than 100 feet per second!". (Am I right? If velocity is acceleration times time, then 3.3 times 32 is 105.6 feet per second). I've timed the grapefruit often, and so have my clients. It's good therapy.

Last week my ex-roommate (the psychokiller of decimated ceramic sculpture fame) found a copy of my brochure and sent me a nasty letter. He says,

> *Merton you jerk! It's not $v = gt$, it's $v = A(e^{at} - 1)/(e^{at} + 1)$! Any physicist worth his beans knows that the terminal speed of a baseball has been measured to be a measly 138 feet/sec. Admittedly, grapefruits are bigger and denser: 10 cm compared to 7.2 cm, and 0.86 g/cm³ versus 0.72g/cm³. Since terminal velocity scales like the square root of the ratio of the diameters for equally dense round objects (duh), and putting this extra mass into the equation, this increases the top speed of a grapefruit to 176 ft/sec. So there's no way your grapefruit could be going 100 ft/sec after just 3.3 seconds. If you keep lying through your teeth about your grapefruits, I'm going to sue your big fat [you get the idea] for false advertising!*

Well, now I'm stuck. I should never have taken the old mallet to the ceramic, that's for sure. And I should have taken more math. What in the heck is A? or a? It's not gravity, and it's not initial velocity. I read in my calculus book that the equation that I used to get 105.6 "assumes no acceleration due to air resistance, which physicists claim is proportional to the square of velocity". Great. Now they tell me. That's just great. Can you help me? I have no idea if my psycho roomie is right about that e junk. Is he? I have no idea how long it'll take a grapefruit to fall as fast as 100 feet per second. Man, I'm about to lose my business, and all because grapefruits can only fall so fast. It's sad, but it's true.

Yours sincerely,
Joe Merton

P.S. As long as you're at it, could you tell me how long a grapefruit would take to pass 150 feet per second? I might just try out this helicopter idea after all.

Solution

Acceleration of a falling grapefruit is $dv/dt = -g + \alpha v^2$, where α is a constant of proportionality. Because terminal velocity means $dv/dt = 0$, we can solve to get $\alpha = g/176^2$. We can plug the crazed ex-roomie's formula into the differential equation and see that it works, subject to fiddling with A and a. Indeed, we see that A is terminal velocity (velocity when time is large), and that $A = -\sqrt{g/\alpha}$ and $a = 2\sqrt{\alpha \times g}$. We can use a calculator to determine the time it takes to get a falling grapefruit up to speed: $t_{v=-100} = 3.52$ seconds and $t_{v=-150} = 6.91$ seconds.

The Case of the Lead Poisoning

Annalisa Crannell

Calculus II

Concepts   Simple system of differential equations

Realism   Real-world application with fabricated storyline. The model used in this project assumes that lead in the bones does not leach back out, which is a simplifying assumption that is nearly, but not entirely accurate.

Reactions   This was the fifth part in a six part series. The students enjoy detective work, and found the differential equations challenging but not impossible.

Credits   Thanks to Dr. Marianne Kelly of Franklin & Marshall Health Services; *Differential Equations* by Borrelli and Coleman

Technology   Scientific calculator

Big City Police Department
Constabulary Avenue
Big City, PU 11235
February 23, 1998

Calculus II Students
Franklin & Marshall College
Lancaster, PA 17604-3003

Dear Calculus Student:

I write to you concerning one Mr. Joseph Merton of Big City, PU. I regret to tell you that Mr. Merton is currently undergoing treatment for a case of lead poisoning, suspected intentional. He requested that I communicate with you regarding the particulars of the situation, and specifically requested that I involve you in our investigation. Under normal circumstances, departmental polices would dictate that I refuse to accede to such requests, as they might interfere with ongoing police inquiries. In this instance, however, I admit to being baffled and will gratefully accept any assistance you might be able to lend, in the role of outside expert consultant, of course.

The facts of the case are these. Approximately 6 months ago, on the 30th of August, Mr. Merton received a set of 6 ceramic earthenware mugs from an unknown source. The card sent with them, which we have in our possession, says merely, "from an old friend. Here. You deserve them." The card was unsigned. Since then, Mr. Merton has made it his daily habit to use these cups for his morning coffee.

Recently, Mr. Merton began to experience abdominal pain and irritability followed by lethargy and slurred speech, typical symptoms of lead poisoning. He was examined by his regular physician, who found him to have a lead-level of 60 mg/dL in his blood. (For a man of Mr. Merton's size this signifies that he has a total of .033 grams of lead in his blood and tissues). His physician immediately admitted Mr. Merton into the Big City Lady of Pity Hospital and simultaneously contacted me.

I had forensics search Mr. Merton's residence and his place of work ("The Citrus Clinic") for possible sources of lead contamination, and they identified the aforementioned mugs. I have since been made aware that many ceramic glazes traditionally used to contain lead, although modern regulations and protections have virtually eliminated such glazes from commercial products. Forensics estimates that these particular mugs leech 7950 μg (or 0.00795 g) lead into a cup of coffee, and concludes from this that they were not mass-manufactured. These mugs were handmade.

It is Mr. Merton's unsubstantiated opinion that these mugs were made by a former college roommate of his, one Mr. Jack Phaze. It seems that these two men had a falling-out some years earlier, and Merton alleges that Phaze harbors a vendetta over this incident. We detained Mr. Phaze briefly for questioning; he denies making the mugs, although he admits to having used lead-based glazes in his ceramic sculptures. I must admit that he struck me as a highly unsavory character, and I doubt his veracity.

Mr. Phaze did raise an interesting point, however. He claims that the quantities of lead involved would be insufficient to raise Mr. Merton's blood lead-levels to their current level, because the human body eliminates lead in proportion to the amount present. His point is that there must be other sources of lead that our Forensics department has failed to discover.

Mr. Merton's physician confirmed this statement, or rather, part of it. She explained that when an adult subject ingests lead, only 15% is absorbed, and this goes directly into the blood and tissues. From there some 0.39% per day is transferred into the skeleton, and some 3.22% per day is eliminated through urine, feces, and sweat. Although some amount of lead deposited in the bones leeches back into the bloodstream, the amount of this transference is negligible unless the subject takes medication.

She assures me that it is the lead deposited in the bones that is deadly, because it is so difficult to remove. She also explained that the lead count on Mr. Merton does not take into account the amount of lead in his bones—indeed, she cannot determine this amount precisely without expensive diagnostic tests.

Mr. Merton believes that you can succeed where others have failed. Even if we succeed in tying Phaze to the mugs, we have no evidence that these alone are sufficient to account for Mr. Merton's elevated lead levels. And Mr. Merton is somehow convinced that you can determine the lead level in his bones, even though his physician cannot. He says, and I quote:

"Tell them that the change in bone level is the blood level, and that the change in blood level is intake minus output. They can solve it from there."

I sincerely hope that he is right.

Sincerely,
Officer Sonia Kovalevskia

Solution

We assume, although the letter does not explicitly say so, that Mr. Merton drinks exactly one cup of coffee each morning. If B is the amount of lead in the blood in μg, then $dB/dt = (.15)7950 - (.0039 + .0322)B$, which is (intake) $-$ (output). Here t is measured in days, rather than months. This differential equation can be easily solved on its own, and we discover that the lead in the mugs does exactly account for Merton's elevated lead levels. If S is the amount of lead in the skeleton, then $dS/dt = .0039B$, which we can solve by knowing the equation for B.

The Case of Darlene's Rose

Annalisa Crannell

Calculus II

Concepts   Graphing and integration in polar coordinates

Realism   Completely fabricated

Reactions   This is the last in a six-part series. I had hoped students would get very experimental with creating polar graphs; their attempts were more modest than I might have supposed.

Technology   Graphing calculator or software with polar graphing abilities

The Citrus Clinic
PO Box 2358
Big City, PU 11235
March 25, 1998

Calculus II Students
Franklin & Marshall College
Lancaster, PA 17604-3003

Dear Calculus Student:

You must be the Lone Math Ranger, the way you swooped in and solved those nasty equations! I owe you my very life, and I won't forget it. I don't know if the unstoppable Officer Kovaleskia told you or not, but they found Phaze's fingerprints on the note that was sent with the mugs (well-preserved under a piece of tape—what a doofus). That evidence, together with your incredible mathematical savvy, has sealed the prosecution's case against him. Phaze is in the hoosegow now! Who was that masked mathematician?!?

My doctor here at the Big City Lady of Pity Hospital is wonderful, too: she's Jonas Salk and Florence Nightingale rolled up into one. She says that the information you passed along about my bones made all the difference in my treatment. She's spent a lot of time working with me, and I feel so much better now—and the reason for my good spirits isn't entirely medical, I must admit. Dr. Clementine (Darlene, to her friends) is a real winner. She's smart, she's athletic, she's pretty—and she even likes grapefruits! Her only blemish, if I dare call it that, is that she lacks depth perception. Still, she's managed to overcome that handicap better than any depth-perception-challenged person I know. I can't think of anyone I'd rather share the rest of my life with, and I'm going to do my best to convince her to feel the same way about me.

So since you're such a studly math-is-no-problem kind of a hero, maybe you can help me one last time. Darlene has been on a real calculus kick ever since some math students showed her track team how to win the 400-meter relay. Her teammate Betty lost her heart to a guy who was good at derivatives and poetry both, and Darlene says that she too is saving her love for the man who can solve her problem challenge. (It sounds like a princess setting a task for a knight, but I'd do anything she would tell me to.)

She has a white-gold medallion in the shape of a circle, exactly 2 cm across. She wants to inlay in this medallion a flower made of yellow gold—but not just any flower: a rose curve. And she needs to know how much yellow gold she'll need. So I tried: a 3-petal rose curve $r = \sin(3\theta)$ has area $\pi/4$ cm^2. (I got this by calculating that the area of each of its petals is $1/2 \int_0^{\pi/3} r^2 d\theta$). The area of a 5-petal rose curve $r = \sin(5\theta)$ is also $\pi/4$ cm^2. But the area of a 4-petal rose curve $r = \sin(2\theta)$ is $\pi/2$ cm^2.

I proudly explained this to my darling Darlene Clementine, but she just laughed: "So what's the area of a 6-petal rose curve?" I shot back, "$\pi/2$ cm^2! ... wait ... the curve $r = \sin(6\theta)$ has 12 petals and $r = \sin(3\theta)$ has 3 petals ... there is no such thing as a 6-petal rose curve!"

She laughed that silvery laugh again and said, "Check out $r = \sin(3\theta/2)$." Lo and behold—I think she may be right! But I have no idea how to get the area of the complicated picture I came up with. I'm hoping you can help me!

Even if I come up with the area of this rose curve, I know she'll come back with more complicated ones. Can you pick a beautiful rose curve, one that would be worthy of my darling Darlene Clementine, and tell me how to find how much gold we'll need for the inlay?

I can just see the Marquis sign now: "Grateful cowpoke gets the girl as mathematician rides off into the sunset." This one last favor, Lone Math Ranger, is all I'm asking of you.

Your humble,
Joe Merton

Solution

The curve $r = \sin(3\theta/2)$ does have six petals, provided we allow θ to vary from 0 to 4π. However, as a calculator shows us, there is overlap between the petals which we shouldn't count twice. A suitable range for integration would be something like from $\theta = \pi/6$ (the first point of intersection) to $\theta = \pi/2$ (the second point of intersection), and then multiply this area by 6 petals.

The Case of the Set-Up Cyclist

Annalisa Crannell

Calculus I

Concepts   Average velocity versus averages of velocities

Realism   Completely fabricated

Reactions   Students could prove that Billy was innocent by determining the time she spent on his bicycle, but they had difficulty refuting the (incorrect) argument of the undercover officer.

Credits   This is an adaptation of one of my favorite Lewis Carroll problems, found in "Mathematical recreations of Lewis Carroll" by Lewis Carroll, Dover Publications, New York, (1958).

Technology   None required

Larry's Law Firm
1729 Easy Street
Big City, PU 11235
September 13, 1993

Calculus Students
Franklin & Marshall College
Lancaster, PA 17604-3003

Dear Calculus Students:

I'm a new lawyer at a prestigious law firm here in Big City, PU. Our firm was recently engaged to argue the speeding ticket of one Billy B. Bartholomew, owner of the Bartholomew Bicycle Barn. Since speeding tickets are, quite frankly, low on the firm's list of publicity-generating cases, it was assigned to the new-comer (i.e., me). It's my first case and, even though it's not a life-and-death matter, I'd still like to do a good job. The problem is, I've gotten in over my head with the mathematics involved, and so I was hoping that you could help me. It was your intrepid and enterprising professor, Dr. Crannell, who referred me to you.

Billy lives in Wheeling, PU, a suburb of Big City, where she runs her Bicycle Barn. Several Fridays ago, she peddled over the hills to a nearby town, Speedwell Valley, to give a demonstration for some bicycle enthusiasts. (She did not know at the time that one of these was an undercover police officer). Now, Speedwell Valley is a conservative town, with some obscure and out-dated laws on their books. A former mayor of Speedwell Valley, Mayor Gotcher, used to own the trolley that ran from Speedwell Valley through the hills to Wheeling and then back again. To protect his financial interests, he passed a law stating that anyone making the trip from Wheeling to Speedwell Valley faster than his trolley would be fined a "speeding" ticket of no more than $200.

On the day that Billy rode to Speedwell Valley, she had used a speedometer to record her speed. For the first 5 miles out of Wheeling, which were flat, she traveled at a constant rate of 25 miles per hour. The next 5 miles were up-hill, and so she slowed down then to 15 miles per hour until she reached the top of the hill. Then she dashed down-hill for 5 miles to her destination, traveling this last leg at 35 miles per hour. Her presentation used the recorded data from her own trip to show how cyclists can work on maintaining a constant pace (which Billy did admirably, everybody agreed).

However, as Billy was leaving, the police officer confronted her and charged her with exceeding the Gotcher speed limit. Since she traveled at an average of 25 miles per hour for the whole trip (averaging the three speeds), he points out that it must have taken her a mere 36 minutes to cycle from her home to the demonstration, and therefore she traveled faster than the trolley (which took 40 minutes).

Billy swears that it took her just above 40 minutes to make the trip, but she also claims that her bicycling data is accurate. Now I'm stuck: can Billy be telling the truth? But if so, how? And if not, how do I defend her against these accusations?

This is why I'm asking you for help—in hopes that you can answer these questions for me. If there is any way that you can find an answer to this dilemma, I'd be incredibly grateful. I would appreciate an answer as soon as possible, but certainly no later than September 30, as our preliminary hearing is the day after.

Yours sincerely,
E. Noether, Attorney
Larry's Law Firm

Solution

The averages of the velocities is not the same as the average velocity. She spent much longer going up-hill at 15 mph than down-hill at 35. Indeed, she spent 40.57 minutes on her trip, and her average velocity was a mere (!) 22.18 mph.

The Case of the Crushed Clown

Annalisa Crannell

Calculus II

Concepts   Parametric equations; projectile motion

Realism   Completely fabricated

Reactions   How do you kill someone with parametric equations? The students seemed to enjoy the murder mystery and to have a "real world application" for projectile motion.

Technology   None required

Police Department
Constabulary Avenue
Big City, PU 11235
November 6, 1992

Calculus II Students
Franklin & Marshall College
Lancaster, PA 17604-3003

Dear Calculus Student:

Our police force has run up against a difficult problem, and we were hoping that we could turn to you for assistance. Your very gracious and upstanding professor, Dr. Crannell, to whom I am going to dedicate my life, told me that you might be able to help.

Our problem is this: a clown from a local circus troupe was shot out of a cannon and into a canyon and subsequently died. We believe that it was murder—in fact, we have a suspect—but we can't prove that it's not a suicide.

The facts of the case are these: Bobo the clown has been performing in the Upsy-Daisy Traveling PU Circus for about 2 years. The Upsy-Daisy is centered in the mountains bordering Big City; they prefer practicing in high altitudes where the air is rarer. Bobo performs a variety of antic acts for the circus, including one in which he is fired out of a cannon and parachutes into the local canyon, Absolutely Gorges. Their next big show was to be November 8, and at the time of the accident they were in the midst of rehearsals.

On the afternoon of November 1, Bobo climbed into the cannon for reasons that we are still unclear about. According to eyewitness accounts, the only living beings within close range were the owner of the circus, Rick Rasterdly, and five African elephants. Somehow the cannon was fired, and Bobo, who was not wearing a parachute at the time, was shot clear across the gorge, hitting the opposite wall some 310 meters down.

It surprised me, to be quite frank, that Bobo could fly all the way that he did. I don't know how far it is across the canyon (measuring down was easier than measuring across), but it must be at least 200 meters. The reason I'm so surprised is that the cannon itself is not very large: when one end is on the ground, the other end is only 3 feet in the air, and from base to mouth it's only 5 feet long. Nonetheless, it can fire off something as large as a human being at a speed of 30 meters per second.

Our main line of questioning has been whether Bobo fired the cannon himself or whether Rasterdly set it off instead. It only takes about a half a second to push the button. We have one eye-witness who swears that Mr. Rasterdly hit the ignition button. We have another who swears that Mr. Rasterdly kept his hands to himself. We have witnesses who swear that Rasterdly and Bobo had been arguing about money, and others who claim that they were best friends. We have witnesses who say Bobo was depressed, and others who say he never seemed happier. To tell the truth, we're getting sick of witnesses, and that's why we're turning to you.

Here is Mr. Rasterdly's version of what happened:

> *I was filming the elephants for a rock video. I figure, if MTV likes sex, violence, and raw, bestial emotion, why not use elephants, eh? Bobo—geez, I still can't believe he changed his name legally—it used to be Winifred Dieselblock—well, he was in the cannon, probably checking things out. He often does before a gig, just to make sure everything's working okay. I didn't pay much attention because I had the elephants there; if you had to choose between 5 elephants or a clown you'd [—] well pick the elephants, wouldn't you? So we're getting to the tricky part where they all stand up on their hind legs and put their front legs on each other, and I hear a 'boom!'. I turn around, and there's Winifred sailing off. I kept waiting for his chute to open and thinking it's a [—] fine time for him to pull this, knowing we'll have to send the jeep down to get him; but his chute never opens and he bangs into the wall on the opposite side and then tumbles on down. I can't believe it was suicide, I just can't. It must have been an accident. He pushed that button by mistake.*

The video recorder confirms his story up to a point: it has no sound, but it has a digital clock which leaves an imprint on the film. Just as the elephants were rearing up (15:52:20 on the film), the camera veered and was dropped.

It turns out we have one further ace up our sleeve, and we're keeping this quiet. Bobo's watch had been fixed in town that week and he'd just picked it up the morning of his death. The watchmaker told us it had kept excellent time during the week since she'd repaired it, not losing or gaining a second from her master clock. Bobo was wearing his watch when he died, and it stopped at 3:56:50. We compared the video camera to the watchmaker's clock, and found the former to be 4 minutes and 16 seconds slow.

What we'd like to know from you, if you can possibly help us with this, are the following three questions: does Rasterdly's story hold water, and can we convince the jury that it doesn't? How fast was Bobo going at the time of impact, and was this sufficient to cause death (there's always the outside chance Bobo was murdered in the cannon, and then disposed of to avoid suspicion). And lastly, just to check that your calculations are correct, can you tell us how far across the gorge Bobo flew? (We'll measure this ourselves to cross-check the accuracy of your computations). I would appreciate your answer(s) by November 24, as we'll go to trial the day after.

Yours sincerely,
Officer S. Kovalevskia

Solution

With a name like Rick Rasterdly, how could he not have done it? If we ignore air resistance (in the 'rare air'), Bobo's position vector had the form $(24t, 18t - 4.9t^2)$. Since Bobo landed 310 meters down, we can figure out that his perilous flight took 10 seconds. This contradicts Rasterdly's statement (which would have implied that Bobo's trip was 14 seconds long), and destroys his alibi. The canyon is 240 meters across. This assumes that the cannon sits precisely at the edge of the canyon, a fact which is not explicitly stated in the letter.

A Bit of Caustic Rain

Gavin LaRose

Calculus I

Concepts   Finding exponential and sinusoidal functions from given data, solving equations involving exponentials.

Realism   Real-world application. The rainfall data is close to actual average rainfall data for Lincoln, NE (Lonlinc, Skanebra), obtained from an almanac, duplicated over three years with noise added in. The HNO_3 concentration is generated by starting with a rainfall pH of a reasonable 5.2, with a 7% increase per year and added noise.

Reactions   I very much liked this project, because it's particularly well grounded in real data and apparent applicability. Students had the most trouble establishing the relationship between the concentration of acid in the rain and the amount of acid deposited on the test area. Many decided (correctly) that they needed only consider the maximum in the rainfall, ignoring the sinusoidal character of the amount of precipitation. There is one anomalous data point which caused some students some consternation.

Technology   Graphing calculator or computer algebra system

Lonlinc CPE
the Lonlinc Building
South 9th Avenue
Lonlinc, SK 04685

Independent Mathematical Contractors, Inc.
Suite 2, Strawmarket Business Plaza
Lonlinc, SK 04685

Dear IMC:
As you know, the Lonlinc, Skanebra Council on the Protection of the Environment (CPE) is charged with the evaluation of all things environmental that may have an impact on the continued health of the residents, animal and human, of the thriving city of Lonlinc. Currently boasting a population of 200,000 and growing at an annual rate of 7%,[1] Lonlinc is regarded by some as increasing in size sufficiently fast as to demand appropriate consideration of the environmental impact of such things as the increased traffic attendant with this growth.

One effect of this increase in traffic is an increase in exhaust pollutants in the air, which may lead to the phenomenon known as "acid rain." This occurs when NO and NO_2 from car exhaust combines with atmospheric water to create nitric acid (HNO_3), which makes

---

[1]Which has lead some observers to speculate that Lonlinc's population will exceed 1,000,000 by the year 2023!

precipitation (rain, snow, etc.) acidic. In an effort to determine if this is likely to soon be a problem in the Lonlinc area, we have over the past three years monitored the rate of precipitation and the concentration of HNO_3 in that precipitation, which data appears in Tables 1 and 2 below.

TABLE 1
Monthly Precipitation (mm)

Month	1	2	3	4	5	6
Precipitation	17.3	20.6	60.0	70.5	109.8	114.0
Month	7	8	9	10	11	12
Precipitation	99.9	98.7	101.6	62.3	35.1	26.0
Month	13	14	15	16	17	18
Precipitation	17.7	21.8	59.6	75.0	110.9	108.8
Month	19	20	21	22	23	24
Precipitation	100.4	102.8	96.5	59.3	36.9	25.5
Month	25	26	27	28	29	30
Precipitation	17.3	20.9	56.7	13.4	108.9	105.5
Month	31	32	33	34	35	36
Precipitation	92.8	103.3	98.3	60.9	53.9	25.1

TABLE 2
Average HNO_3 concentration in precipitation (g/ml)

Month	1	2	3	4	5	6
HNO_3 conc.	0.4003	0.4033	0.4060	0.4078	0.4100	0.4124
Month	7	8	9	10	11	12
HNO_3 conc.	0.4136	0.4171	0.4184	0.4214	0.4248	0.4255
Month	13	14	15	16	17	18
HNO_3 conc.	0.4282	0.4307	0.4331	0.4356	0.4380	0.4409
Month	19	20	21	22	23	24
HNO_3 conc.	0.4440	0.4454	0.4490	0.4513	0.4532	0.4567
Month	25	26	27	28	29	30
HNO_3 conc.	0.4587	0.4603	0.4638	0.4658	0.4700	0.4715
Month	31	32	33	34	35	36
HNO_3 conc.	0.4743	0.4772	0.4805	0.4834	0.4852	0.4879

The critical factor in determining the severity of the acid rain is, of course, the amount of HNO_3 that is deposited on the surfaces on which it falls. We therefore need to establish, based on the data we have collected, how the monthly rate of chemical deposition is changing in time, and to predict when (if ever) this is likely to exceed 11.5 g/month on a given 1 cm^2 area.

We appreciate your prompt attention to this matter, and look forward to receiving your final report of 3–5 pages. To assure your success in the endeavor, our department's scientific expert (none other than your fine instructor) will be available to answer any questions that you might have in the course of your investigation. *Said expert will, however, be unavailable to assist with this project over the weekend before the project is due.* You should also plan on meeting with our expert next week to verify your initial progress towards completion of the project. You may at that time also be able to obtain specific instructions or sample reports that may prove useful as you develop your formal response.

Sincerely,
Jack C. Ousteaux
Director, Lonlinc CPE

Solution

The total amount of acid deposited on a 1 cm^2 area is $c(t) \cdot p(t)$, where $c(t)$ is the concentration of acid in the precipitation and $p(t)$ the precipitation, both as functions of time. The functions $c(t) = 0.3980(1.0059)^t$ and $p(t) = 46.8 \sin(\frac{\pi}{6}t - \frac{2\pi}{3}) + 64.1$ should fit the data reasonably well. Solving for when $c(t) \cdot \max(p(t)) = 11.5$ (after adjusting units appropriately, of course) gives $t \approx 166$ months.

A Bit More Caustic Rain

Gavin LaRose

Calculus I

Concepts   Applying the Fundamental Theorem of Calculus to find the total change of a function given its rate. This amounts to finding the integral (Riemann sum) of the rate function, which must be determined from the data given.

Realism   Real-world application. The rainfall data is close to actual average rainfall data for Lincoln, NE (Lonlinc, Skanebra), obtained from an almanac, duplicated over three years with noise added in. The HNO_3 concentration is generated by starting with a rainfall pH of a reasonable 5.2, with a 7% increase per year and added noise.

Reactions   Assigned after the project "A Bit of Caustic Rain," most students didn't have too much trouble figuring out what the data meant. The most difficult steps were for them to determine how to mathematically find the accumulation of acid on a specific area for a fixed and then longer time.

Technology   Graphing calculator or computer algebra system

Lonlinc CPE
the Lonlinc Building
South 9th Avenue
Lonlinc, SK 04685

Independent Mathematical Contractors, Inc.
Suite 2, Strawmarket Business Plaza
Lonlinc, SK 04685

Dear IMC:
Our great thanks for your prompt and useful work on the matter of the acid rain that we are hoping not to suffer from in Lonlinc. We have forwarded your work to the various agencies of import to which we inevitably report, and have received numerous accolades for its comprehensive assessment of the problem. However, we have also received a reply from the transportation division of the Lonlinc bureaucracy. They are in the process of installing high-gain solar panels to power a number of traffic direction devices about the city, and the surfaces of the panels will eventually become pitted and therefore less clear as they are exposed to the acidity of the precipitation. According to their estimates, it would take exposure to an amount of Nitric Acid (HNO_3) equal to 800 grams for this to have a significant impact on the solar panels' electrical yield.

   We therefore need to know, based on the data for monthly precipitation and HNO_3 concentration reproduced in tables 1 and 2 below, at what point the total acidic exposure to which a 1 cm^2 section of a solar panel will have been subjected will equal or exceed 800 g of HNO_3.

It is with great pleasure that we anticipate your final report of 3–6 pages on this matter. Owing to the success of the secondary consulting arrangement implemented with your previous project work, we have again made our scientific expert (who continues to be none other than your fine instructor) available to answer any questions that you might have in the course of your investigation. *Again, this estimable expert will unfortunately be unavailable to assist on this project over the weekend before its due date.* You should, with your project team, also plan on meeting with this expert sometime next week.

Sincerely,
Jack C. Ousteaux
Director, Lonlinc CPE

TABLE 1
Monthly Precipitation (mm)

Month	1	2	3	4	5	6
Precipitation	17.3	20.6	60.0	70.5	109.8	114.0
Month	7	8	9	10	11	12
Precipitation	99.9	98.7	101.6	62.3	35.1	26.0
Month	13	14	15	16	17	18
Precipitation	17.7	21.8	59.6	75.0	110.9	108.8
Month	19	20	21	22	23	24
Precipitation	100.4	102.8	96.5	59.3	36.9	25.5
Month	25	26	27	28	29	30
Precipitation	17.3	20.9	56.7	13.4	108.9	105.5
Month	31	32	33	34	35	36
Precipitation	92.8	103.3	98.3	60.9	53.9	25.1

TABLE 2
Average HNO_3 concentration in precipitation (g/ml)

Month	1	2	3	4	5	6
HNO_3 conc.	0.4003	0.4033	0.4060	0.4078	0.4100	0.4124
Month	7	8	9	10	11	12
HNO_3 conc.	0.4136	0.4171	0.4184	0.4214	0.4248	0.4255
Month	13	14	15	16	17	18
HNO_3 conc.	0.4282	0.4307	0.4331	0.4356	0.4380	0.4409
Month	19	20	21	22	23	24
HNO_3 conc.	0.4440	0.4454	0.4490	0.4513	0.4532	0.4567
Month	25	26	27	28	29	30
HNO_3 conc.	0.4587	0.4603	0.4638	0.4658	0.4700	0.4715
Month	31	32	33	34	35	36
HNO_3 conc.	0.4743	0.4772	0.4805	0.4834	0.4852	0.4879

Solution

Given the functions obtained in the "A Bit of Caustic Rain" project, the total deposition over time is given to a time t by $\int_0^t c(t) \cdot p(t)\, dt$, where $c(t)$ is the concentration of acid in the precipitation and $p(t)$ the precipitation, both as functions of time. The only subtlety is that this works only because we specified a 1 cm^2 area—otherwise we would need some multiplier for the area being considered. The functions $c(t) = 0.3980(1.0059)^t$ and $p(t) = 46.8\sin(\frac{\pi}{6}t - \frac{2\pi}{3}) + 6.41$ should fit the data reasonably well. A time $t \approx 176.82$ (months) gives a deposition of about 800g.

Profitable Production

Gavin LaRose

Calculus I

Concepts   Function optimization using derivatives.

Realism   Real-world with fabricated data. Both the author and a member of his school's business department concluded that the model used here is not unreasonable for the situation described, but it's probably not something that anyone actually uses.

Reactions   For the students the really formidable part of this project is its lack of numbers. Students responded well to the suggestion that they try to work out the problem by plugging in some values for the different parameters before working the problem in greater generality. The other major hurdle is deriving the equations for the profit and constraint equations, which are explained in words rather than given to them explicitly. Finally, there is a little bit of algebra involved in the solution.

Credits   This is inspired by linear programming problems, with the addition of the non-linearity to make it nontrivial.

Technology   None required

Chemproc, Inc.
20000 Ryan-ears Blvd.
Lonlinc, SK 04685

Independent Mathematical Contractors, Inc.
Suite 2, Strawmarket Business Plaza
Lonlinc, SK 04685

Dear IMC:

As you know, Chemproc, Inc. is a premier manufacturer and reprocessor of chemicals and chemical waste. Further, as has been recently highlighted on the local news, we are in the process of expanding our Lonlinc, Skanebra chemical manufacturing plant to add two or three new products to our elegantly packaged and painstakingly marketed product line—which will result in our hiring at least *four* local workers. Our engineers have, however, determined that for these products to be successful in today's competitive market we must be careful to obtain the absolute highest profit possible.

We are, for reasons of corporate secrecy, unable to reveal the exact names of the products that we will be producing, and therefore refer to them herein as products X and Y (for the case in which two products are produced), respectively. We will be manufacturing x and y units of these per day, and expect to realize a profit of a and b dollars per unit on each of the products (again, we are unable to divulge the actual values determined by our marketing department).

The actual number of units of these that we can produce is, however, limited by the number of person-hours that the workers have available. This will be L person-hours. To manufacture one unit of X requires c person-hours, and the hours required for more units is proportional to the number of units of X produced. For the second product, however, economies of scale are much more pronounced, so that the number of person-hours required to produce y units of it is proportional to y^p (where $0 < p < 1$), with a constant of proportionality d.

It is imperative that we determine the best production strategy for the manufacture of these two products. We are additionally interested in the case in which we manufacture three products, X, Y, and Z, and hope that you will find it possible also to investigate this possibility. In this case the economy of scale indicated above is only applicable to the third product, while the time required for the other two is linear in the number of units produced. The profit on these is a_1, a_2, and a_3 dollars per unit. For this case we have in addition to the time constraint noted above the additional constraint that both of X and Y require a primary reagent of which we are only able to allocate M units per day. Every unit of X requires c_1 units for this reagent, while each unit of Y requires c_2 units. We would be very interested in your determination of the optimum production strategy in this case as well.

As specified in your contract, your final report should be 3–6 typewritten pages. If you should find in the course of your investigation that you have questions regarding this project, you are to contact a most estimable expert, your fine instructor, who is our consulting scientist (whose services we recently obtained by tripling to four figures the salary they offered by the public sector). We regret, however, that owing to other responsibilities you should not expect to obtain assistance in the course of the weekend prior to the project's due date.

It is difficult to indicate the eagerness with which we await your results.

Yours most sincerely,
E. Idu Pont
President, Chemproc, Inc.

Solution

The equation to optimize is in this case $P = ax + by$, with the constraint that $L = cx + dy^p$ is fixed. Solving for x in the latter and substituting into the former,

$$P = \frac{a}{c}L + by - \frac{ad}{L}y^p.$$

Differentiating, setting to zero, and solving (remembering that $0 < p < 1$) gives

$$y = \left(\frac{pad}{Lb}\right)^{1/(1-p)}.$$

This is one case where the second derivative test is much nicer than the first:

$$P'' = \frac{(1-p)pad}{L}y^{p-2} > 0,$$

so this is actually a minimum(!), and the maximum profit is given when

$$x = x_0 = \frac{L}{c} \quad \text{or} \quad y = y_0 = \left(\frac{L}{d}\right)^{1/p},$$

according to whichever of

$$P(x_0, 0) = \frac{aL}{c} \quad \text{or} \quad P(0, y_0) = b\left(\frac{L}{d}\right)^{1/p}$$

is greater. The solution in the second case proceeds similarly.

The Legality of the Stock Market

Gavin LaRose

Calculus I

Concepts   Manipulating piecewise-continuous functions. Finding functions from a given graph, in particular, estimating an exponential growth rate from messy data. Solving exponential equations using logarithms.

Realism   Real-world with fabricated data. The graph given is completely fabricated, though some effort was made to introduce "dives" in the stock value at approximately the correct dates.

Reactions   Students had to think fairly hard to determine the relationship between the data given and what they were trying to find. They had some trouble with the "messiness" of the data and believing that it was reasonable to fit an exponential through it. Students also had difficulty considering the function giving the value of the stock in terms of time as a piecewise function. They liked the application of math to the stock market.

Technology   Graphing calculator

Hangemhi, Inc.
Suite 101, Boldledge Business Park
Lonlinc, SK 04685

Independent Mathematical Contractors, Inc.
Suite 2, Strawmarket Business Plaza
Lonlinc, SK 04685

Dear IMC:

As recognized legal experts here in Lonlinc, and as a result of our reputation in criminal law, we have recently been contacted by the family of a sadly departed former client to execute (no pun intended) their will.

Our difficulty arises in the division of the largest portion of the deceased's wealth, which is in stocks and which was invested approximately 50 years ago. However, for reasons of which we are unwilling to speculate, the exact amount invested was not recorded at that time. We do know that the money was invested in a number of different stocks (much in the way mutual funds are run today), the exact companies of which changed at the whim of the owner but which generally stayed in one sector of the market. We also know that on release from prison in 1970 the (now) deceased invested an additional $100,000, with a like amount also being invested in 1980 immediately before a subsequent reincarceration. The general performance of this sector of the market over this time period is shown in Figure 1.

The total value of the stocks is currently $2.4 million. In order to execute the will, however, we need to know two things: first, a model for the value of the stocks in the

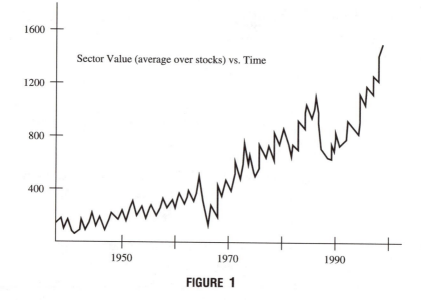

FIGURE 1

portfolio over time, and second, the amount of the original investment. To facilitate your work we have arranged that your instructor, an eccentric mathematician in the area who appears to have all sorts of contacts, will be available to assist you with any technical questions you may have in the process of resolving this issue. Furthermore, as we will be working on the case in the intervening time, we need summaries of your progress by the end of each of the weeks in which you are working on the project, which may be submitted orally to your instructor. You will understand that this is a weighty matter, and it is on account of this that we must reserve the right to terminate your contract without notice if any of these deadlines are not observed. Your instructor is also able to provide you with example formats which may assist in the formulation of your 3–5 page final report.

We look forward to hearing from you.

Sincerely,
Claire N. "C.D." Arro
Partner, Hangemhi, Inc.

Solution

The objective in this case is to find an equation modeling the performance of 'this sector' of the stock market, and then use it to back-calculate the value of the stocks at the time of the original investment, which we are told is generally in the same market sector. Taking time $t = 0$ to be 1940, and assuming that the origin of the given graph is at (0,0), we can estimate the points (0,150) and (60,1500) as being on the graph. If the curve between these points is exponential, it is $s(t) = 150e^{t \ln(10)/60}$. We know that the current value of the investment is approximately 2.4×10^6, so the value in 1980 (immediately after the second reinvestment) was about $(2.4 \times 10^6)(e^{-20 \ln(10)/60}) \approx 1.1 \times 10^6$. Less 100,000, this is about 1,000,000. Thus in 1970 the investment was worth $(1 \times 10^6)(e^{-10\ln(10)/60}) \approx 6.8 \times 10^5$. Again, less 100,000, about 580,000. Finally, 20 years before that we'd have about $(580{,}000)(e^{-20\ln(10)/60}) \approx 270{,}000$.

An Oily Mess

Gavin LaRose

Calculus I

Concepts   Finding an exponential function from inexact data. Calculating the total change in a function from its rate of change using the Fundamental Theorem of Calculus with left- and right-hand sums.

Realism   Real-world with fabricated data. It seems reasonable that oil-field yields would be exponentially decreasing. All of the numbers in this particular project are fabricated, though.

Reactions   There are a number of tricky points in this project. First, we say the oil field is "for all intents and purposes" not yielding oil now, but we also say that the yield is exponential—which isn't ever zero. Students therefore have to come up with some functional definition of "for all intents and purposes." Second, we don't specify the initial yield, which is a common starting point for the calculation of an exponential function. This requires that students determine the function less directly, by finding the rate from the information given and then integrating to find the total yield (which we know) and thereby solving for the initial rate. And third, there is the general problem of understanding the Fundamental Theorem of Calculus.

Technology   None required, although it is nice to graph the functions.

The Dran Corp.
1 Dran Rd.
Gos Andies, IF 80999

Independent Mathematical Contractors, Inc.
Suite 2, Strawmarket Business Plaza
Lonlinc, SK 04685

Dear IMC:
As you know, the Dran Corporation is a large scale think-tank that analyzes all manner of real-world situations for our clients, much in the same manner as your company operates. Unfortunately, we have found it expedient to subcontract a number of our analyses on account of the recent elopement of a brilliant mathematician with our (we like to emphasize) Ex-CEO.

In this case, we are investigating the rate of oil extraction from a now-defunct oil field to gain insight on the progression of returns from such fields as they mature. We know for this field that it is now for all intents and purposes not yielding oil, and that since it was first drilled in 1967 it has yielded approximately 55 billion barrels of oil. In addition, the mathematician (*c.f.* above) we previously had working on the problem concluded that the rate of oil extraction from a field of this type is an exponential function of time.

Given this information, we need you to obtain a description of the yield (in barrels per day) of this oil field at arbitrary times in its useful life, as well as a similar description of the amount of oil extracted from the field. As is our custom, we have established a liaison in the local scientific community whom you should contact with any questions you may have in completing this project. This is your enthusiastic instructor, who begged to be allowed to complete the project independently. Unfortunately, your instructor's fee was too high, and we have therefore opted instead for a consulting retainer. You may ask said instructor any questions you may have about this project. Your final 3–5 page report should be typewritten and carefully composed. Please be advised that, owing to the large amount of oil-money that is riding on our obtaining a successful solution to this problem, any failure to meet project deadlines will result in a significant penalty to your grade.

We look forward to hearing from you.

Sincerely,
W. R. N. Cristof'r
Consultant-in-Chief, Dran Corp.

Solution

If we assume an exponential function for the rate at which oil is extracted, we have $r(t) = Ab^t$. To find b, assume that the rate is some "minimal" fraction of its original value when $t = 30$—I'll use 1.5%. Then we can take $r(30) = 0.015r(0)$, so $b = 0.87$. Hence 55 (= the total yield for the oil field) $= \int_0^{30} r(t)\, dt = A \int_0^{30} (.87)^t\, dt$, so $A = 7.78$. Note that we can do this with left- and right-hand sums as well, and get an error estimate: the error for the average of a left- and right-hand sum is error $= \frac{1}{2}|r(30) - r(0)|\Delta t$. So if we require that the error be less than 1.5% of the total oil extracted, $(0.015)(55) > \frac{1}{2}A(1 - 0.87^{30})\Delta t$, and $\Delta t = .15$ ($n = 200$ subdivisions in the sum) assures that the error is less than 1.5% (provided $A < 11.2$, which it is). A function giving the amount extracted as a function of time is $a(t) = \int_0^t r(t)\, dt$, which may be best represented with a table of values or a graph.

An Illuminating Project

Gavin LaRose

Calculus I

Concepts   Optimization with a vengeance.

Realism   Real-world with fabricated data. The intensity of light does decrease as the square of the distance. I'm not sure how realistic the decrease in intensity at the edges of the light is, but it makes a nicer problem.

Reactions   The hardest parts of this project were determining the equations to solve, and then dealing with the algebra and the fact that it isn't possible to solve the resulting equation analytically. Students struggled both with the sine function for the bulb intensity and writing the distance the light travels in terms of trigonometric functions.

Technology   Graphing calculator or computer algebra system

Welbilt Design and Construction, Inc.
6533 Haybaler Highway
Lonlinc, SK 04685

Independent Mathematical Contractors, Inc.
Suite 2, Strawmarket Business Plaza
Lonlinc, SK 04685

Dear IMC:

As you may know, Welbilt Design and Construction has recently obtained a contract with the Lonlinc Airport Commission to renovate and expand the well-used Lonlinc Airport. Impressed by work your company has done for us in the past, we are again commissioning your services in support of this project.

As a major contractor for parts of this reconstruction, Welbilt is in charge of the new lighting to be installed in the airport. The modern, attractive fixtures to be installed are shown in Figure 1.

As can be seen in the figure, the light emanates from the fixture in a direct manner through an arc of angle φ_m measured up from the vertical. The specifications of the bulbs to be installed in the fixtures indicate that the intensity of the light shed by the bulb is proportional to the sine of the angle from the vertical, φ, reaching zero at either end of the light-emitting aperture. The intensity of the light hitting the floor is obviously also dependent on the distance from the bulb to the point on the floor being considered, varying inversely as the square of the distance from the fixture to that point (e.g., the point P in Figure 1). The intensity of the light reaching P is therefore the product of these two effects.

To properly plan the positioning of these fixtures, we need to know where on the floor under the fixtures the intensity of the direct light is the greatest, and it is to obtain an

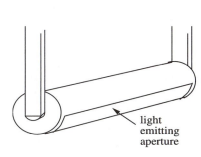

A. Light Fixture, Plan View

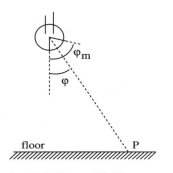

B. Light Fixture, Side View

FIGURE 1
Lighting Fixture

estimate for this that we are contacting you. To facilitate your work, we have arranged with a local mathematical expert and friend of ours, your redoubtable instructor, to serve as a contact for any questions you may have in your work. You should contact this mathematical expert with a report of your progress by the end of each of the weeks in which you are working on the project. Any failure to meet these deadlines will result in an arbitrarily large penalty to your grade.

Sincerely,
Frankyl O. Y. Drite
President, Welbilt Inc.

Solution

As with any optimization problem, this requires finding the function to optimize and then all of the derivative action. In this case, the intensity $I(\varphi) = k$ (bulb effect)(distance effect). The first of these is well represented by the function $\sin(\frac{\pi}{\varphi_m}\varphi)$, while the latter is most easily written also in terms of trigonometric functions as $1/(\text{distance})^2 = \cos^2(\varphi)/H^2$ (where H is the fixture's height above the ground). Thus the intensity as a function of φ is

$$I(\varphi) = \frac{k}{H^2} \sin\left(\frac{\pi}{\varphi_m}\varphi\right) \cos^2(\varphi).$$

Differentiating and setting the derivative to zero gives the requirement that

$$\frac{\pi}{\varphi_m} \cos\left(\frac{\pi}{\varphi_m}\varphi\right) \cos^2(\varphi) - 2\sin\left(\frac{\pi}{\varphi_m}\varphi\right) \cos(\varphi) \sin(\varphi) = 0.$$

Unfortunately, this isn't solvable by hand. However, we can get results for different values of φ_m; e.g., for $\varphi_m = \pi/3$, $\varphi = 0.426$, and a graph of the derivative clearly shows that this is a maximum.

A Question of Law

Gavin LaRose

Differential Equations

Concepts   Solving first-order differential equations models.

Realism   Real-world application

Reactions   The hardest part for the students was formulating the problem and then carrying out the calculations—which are much more difficult if the differential equation is not rewritten with the distance fallen being the independent variable. Relatively few of the students rewrote the equation on their own. Almost all students thought the problem was fun, at least in an abstract sense before having to solve it.

Technology   This is a classic first semester undergraduate ODE problem in disguise.

Technology   None required, though if the equation is solved in terms of t a computer algebra system might be handy for some of the integration.

Hangemhi, Inc.
Suite 101, Boldledge Business Park
Lonlinc, SK 04685

Rigorous Mathematical Contractors (RiMaC), Inc.
Suite 3, Strawmarket Business Plaza
Lonlinc, SK 04685

Dear RiMaC:

In the course of our latest legal entanglement, we are in the position of defending a client who is purported to have leapt (for reasons which, you will clearly understand, we are not in a position to speculate) from a window of a building. So as to obtain all conceivable insight into this possibility (however slight), we are attempting to determine the velocity that our client might have attained in the course of allegedly plummeting from one of the windows in question to the ground below. Unfortunately, owing to the busy nature of the street, we are unable to directly measure the velocity a projectile of the appropriate weight (approximately 180 pounds) would attain in the course of the alleged descent—on account, of course, of the possible impingement on the safety of those who might be found in the street below and resulting legal complications.

   We therefore need your assistance to obtain a mathematical solution to the problem of what the velocity of our client would be if (s)he were to have fallen from a first story window (approximately 5 feet from the ground), second story window (approximately 15 feet from the ground), a third story window, and, for good measure, a fifth story window as well. In that one of our partners is an accomplished sky-diver, we are certain from her experience that the velocity will not exceed the terminal (no pun intended) velocity of 120 miles per hour, but suspect that the velocity of our client will be rather lower than that.

Your physical measurements department has determined for us that the force on a falling body (or, as the case might be, other projectile) may be modeled by $f = m\,g - K\,v^2$, where g is the acceleration due to gravity, m the mass of the body, and v its velocity, but did not provide a solution to the problem as stated.

As specified in your contract, your report should be submitted in typewritten format. If you should find in the course of your investigation that you have questions regarding this project, you are to contact your estimable instructor, who is moonlighting as our firm's consulting scientist (and is, unfortunately, only working for us part-time and hence is unable to resolve this problem for us directly) with the other members of your investigative team.

We look forward to seeing your finished report.

Yours sincerely,
Claire N. "C.D." Arro, Partner
Hangemhi, Inc.

Solution

A differential equation modeling the falling "body" is $mv'(t) = mg - Kv(t)^2$, with initial condition $v(0) = 0$. We can rewrite this in terms of height, noting that $v'(t) = v'(h)h'(t) = v'(h)v(t)$, so that the differential equation becomes $mv'v = mg - Kv^2$, with initial condition $v(h_0) = 0$. This is easily solved with separation of variables to find

$$v(h) = -\sqrt{\frac{mg}{K}(1 - e^{-2K(h_0-h)/m})}.$$

The terminal velocity is the limit of this function as $h \to -\infty$, which allows calculation of K. If the equation is not re-written, the solution for $v(t)$ may still be obtained by separation of variables, but involves partial fractions (or a computer algebra system)—and it also becomes necessary to integrate $v(t)$ to find the distance fallen as a function of time (because this is the only way to explicitly determine when the body hits the ground). In this case

$$v(t) = \sqrt{\frac{mg}{K}} \cdot \frac{\exp(\sqrt{Kgt/m}) - \exp(-\sqrt{Kgt/m})}{\exp(\sqrt{Kgt/m}) + \exp(-\sqrt{Kgt/m})}$$

(a hyperbolic tangent).

A Little Farming of Sorts

Gavin LaRose

Differential Equations

Concepts   Graphical analysis of nonlinear first-order differential equations, interpreting the physical meaning of results, nondimensionalization.

Realism   Real-world with fabricated data. While the model (c.f. credits) is reasonable, I haven't made any effort to justify the choice of parameter values for this particular case.

Reactions   The three steps that caused students difficulty were: First, nondimensionalizing the problem. For many students this was their first experience with this technique, and they found it heavy going. Second, determining the best way to analyze the problem. Some tried direction fields (which are not as useful as phase-planes (phase-lines) here) or repeated numerical solution (which is essentially doomed to failure). Finally, many had trouble articulating what the different terms in the equation physically represented, and thus had trouble indicating why this is a reasonable model.

Credits   The model used here is taken from J.D. Murray's book, *Mathematical Biology* (Springer-Verlag, New York: 1990), where it appears as an exercise in his chapter 1 as a model for a population subject to rapid predation. I haven't bothered to justify why Lonlinc, Skanebra would be subject to such rapid predation, and have surreptitiously avoided the question of how best to actually harvest the fish, but it's a nice model.

Technology   Numerical differential equations solver or computer algebra system; some tool capable of drawing phase-planes (phase-lines) for first-order differential equations

EcoSystems, Inc.
1 EcoSystem Dr.
Leseatt, TG 71986

Rigorous Mathematical Contractors (RiMaC), Inc.
Suite 3, Strawmarket Business Plaza
Lonlinc, SK 04685

Dear RiMaC:

As you may know, EcoSystems, Inc., has an immensely successful series of fish farms that dot the scenic Tonwashing coast a short distance from Leseatt and provide a needed and ecologically sound food source for thousands of satisfied customers who are reached daily by our exclusively electric fleet of bright blue delivery vans. As our hallmark has always been the freshness (not to mention the natural and unsullied nature) of our fish, we have to date unfortunately not been able to expand our distribution to include other parts of the country, such, of course, as Lonlinc.

However, we have recently been offered the opportunity to take over a large lake not far from Lonlinc, which would permit the establishment of a fish farm in that location. Needless to say it is essential that if we approach such an undertaking it be from a position of absolute assurance that it will be able to succeed, and it is for the analysis of a model of the farm that we are approaching your firm.

It is our experience that the reproduction rate of the fish is both proportional to the size of the fish population and limited by the number of fish that the farm can support. Additionally, especially in such a location as Skanebra, we expect predation to be significant. While it should be possible to restrict this to a reasonable level, predation will produce a measurable effect on the fish population whenever there are significant numbers of fish present.

To model this situation, an outside consultant proposed the model

$$\frac{dN}{dt} = RN\left(1 - \frac{N}{K}\right) - P(1 - e^{-N^2/\epsilon A^2}). \tag{1}$$

Her contract, however, was from several years ago, and required only that she come up with this (and a number of other) model(s)—which we have only recently come to require. Owing to a clerical error, much of the explanation that was associated with this particular model was misplaced, though we understand that N is the number of fish, R, K, P and A are constants, and ϵ is a parameter in magnitude very much less than 1. A short distance further in the limited documentation that we have on the model, the consultant concludes that "by substituting $t = \alpha\tau$ and $N = \beta u$ into this equation, it is possible to choose α and β to simplify it to the form

$$\frac{du}{d\tau} = ru\left(1 - \frac{u}{q}\right) - (1 - e^{-u^2/\epsilon}), \tag{2}$$

where r and q are, again, constants." In this equation, (i) ϵ is small; (ii) q is close to 1; and (iii) r is related to the reproduction rate of the fish, which we can control through feeding policy—we expect that it may range from $r = 1$ to perhaps $r = 30$.

Your contract states that you are to justify and analyze the model proposed in equation (2), covering in particular the following issues:

- the derivation of equation (2) from equation (1);

- an analysis of the validity of equation (2) as a model for the fish population in a fish farm;

- an analysis of whether, based on model equation (2), we may expect a stable fish population from which harvesting could take place; and, if so,

- an analysis of how large an initial population of fish will be required to obtain this stable population, and the length of time required for the stable population to be established.

If you should find in the course of your investigation that you have questions regarding this project, you are to contact, as a team, your inimitable instructor, who has agreed to expend copious free time to serve as a mathematical consultant for this project. You should also contact said consultant by the end of your first full week of work on the project (also as a team) to report on your progress.

Yours sincerely,
"Chuck" R.D. Arwin
President, EcoSystems, Inc.

Solution

The first step in this solution is the indicated nondimensionalization. Letting $N = \beta u$ and $t = \alpha \tau$ and simplifying, the equation becomes

$$\frac{du}{d\tau} = R\alpha u \left(1 - u\frac{\beta}{K}\right) - \frac{P\alpha}{\beta}\left(1 - e^{-\beta^2 u^2 / \epsilon A^2}\right).$$

Thus $r = AR/P$, $q = K/A$, $\alpha = A/P$, and $\beta = A$. By examining each term in the equation to determine its effect on the population growth rate we can conclude that this is a reasonable model. To determine what the equation tells us, it is most convenient to look at the phase-lines. Because q is "close to 1," we can take $q = 1$, and because ϵ is "small," we could choose $\epsilon = .001$, or something similar. Then the only parameter to consider is r; a good range of values to consider is $1 \leq r \leq 30$. For small r ($r < 5$ or so), there is no stable equilibrium solution other than $u = 0$, and the fish population will die out. For larger values of r there are two or three equilibria. A solution using a numerical solver is the best way to obtain an answer to the questions posed in the last requested item.

Defending Freedom, or Something Like That

Gavin LaRose

Differential Equations

Concepts Solution and analysis of systems of nonlinear equations.

Realism Real-world with fabricated data. While the physics of following a projectile in space used here are reasonable, the actual values used in the problem and the use of this type of pursuit model in this context are probably unrealistic.

Reactions This is not an easy problem for students. The most difficult parts are deriving the equations for the rocket's motion, obtaining meaningful numerical solutions to the equations, and dealing with the numerical difficulties that are possible in the attempted solution—*viz.*, that the positions of the rocket and asteroid may coincide, that when the rocket and asteroid's speeds are very similar the solution becomes problematic, and that attempting to numerically solve the equations after solving for $x'(t)$ or $y'(t)$ requires solving a problem with a square root, which has an ambiguous sign. Students did like the application, and a number were quite happy to point out that the guidance system suggested here is a very poor one. The realization that the missile could hit the space station was greeted with appreciative groans.

Credits This is a classic pursuit problem repackaged in an appealing space-age package.

Technology Numerical differential equations solver or computer algebra system

Utoff A.F. Base
1 Piecemeal Dr.
Haoma, SK
13681-0050

Rigorous Mathematical Contractors (RiMaC), Inc.
Suite 3, Strawmarket Business Plaza
Lonlinc, SK 04685

Dear RiMaC:
As you undoubtedly know, the national space agency is in the process of developing a space station soon to be built in orbit. We have been, on account of our outstanding success in the development of and use of such defensive works as the Otpatri missile system, commissioned to assess the feasibility of designing a system to protect the space station from space asteroids and debris (of which there is an increasing amount in orbit!).

Our current plan is to put two defense units in orbit around the space station Omfreed, positioned as shown in Figure 1, 180° apart. Then, if an asteroid (A) approaches the space station (S), the closest defense unit (D) will fire a rocket to remove all chance of harm to the station.

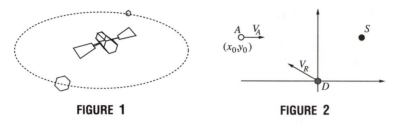

FIGURE 1   FIGURE 2

To keep the required resources to a minimum, we are first considering a very simple guidance system for the defense rocket in which the rocket maintains an orientation which always points towards the asteroid. As shown in Figure 2, this is true at the point at which the asteroid is first found to be a problem ((x_0, y_0) in Figure 2), and will be true at all points thereafter.[1]

To keep things specific, we would like to consider the situation shown in Figure 2, where $V_A = 5$ miles/sec, $V_R = 20$ miles/sec, and the x and y distances from the asteroid to the defense unit are 15 miles and 5 miles, respectively. Suppose that the space station S is one mile (in the x direction) from the defense unit and in line with the asteroid. In this case, will the defense unit succeed in protecting the space station? Does it matter whether V_R is much larger than V_A?

If you have questions regarding your investigation, please feel free to contact your instructor, a technical expert in many fields, with whom we have an affiliation for this project through the space agency. Please note that you should in any event contact this expert with an update on your progress at the end of each full week in which you are working on the project. We look forward to seeing the results of your work.

Yours sincerely,
Lieutenant General Rick N. Backer
Commander, Rocket Tech Division

Solution

This first requires solving for the differential equations for the rocket's position. We consider only two dimensions. The fact that the rocket is always pointing towards the asteroid means that the tangent to its path will always be the line connecting the instantaneous positions of the rocket and the asteroid. If the rocket's position is ($x(t)$, $y(t)$), this requirement is

$$\frac{dy}{dx} = \frac{y(t) - 5}{x(t) - (-15 + v_a t)}.$$

Because $dy/dx = (dy/dt)/(dx/dt)$, this is

$$(x + 15 - v_a t)\frac{dy}{dt} = (y - 5)\frac{dx}{dt}.$$

---

[1] Your tremendously useful physical measurements department has suggested that this will require first making a statement about how the slope, dy/dx, of the rocket's path is related to the x and y positions of the rocket and the asteroid, and then rewriting this in terms of the x and y velocities of the rocket, $x'(t)$ and $y'(t)$.

A second equation is obtained from the requirement of a constant velocity,

$$\left(\frac{dx}{dt}\right)^2 + \left(\frac{dy}{dt}\right)^2 = v_r^2,$$

and the initial conditions are $x(0) = y(0) = 0$. Note that this system gives two solutions—one with the path of the rocket always pointing towards the asteroid, but with the rocket moving away from the asteroid. The remainder of the project involves looking at numerical solutions to this system to determine their behavior. We can in any case verify that an interception has taken place by numerically solving for a time at which the rocket is at the same location as the asteroid.

The results of this investigation may be summarized as follows: if $v_r \gg v_a$, the rocket intercepts the asteroid without problems. If $v_r \sim v_a$, it may or may not be possible to make the interception, and is numerically more difficult to determine. As a specific case, if $v_r = 20$ and $v_a = 18$ an interception does occur at approximately $t = 0.6$ seconds. Note that if v_a is slightly larger than v_r it is possible for both the asteroid and the rocket to hit the space station—in the words of one student, "this is very bad."

A Jump and a Jerk

Gavin LaRose

Differential Equations

Concepts   Solving piecewise first-order differential equations, mostly numerically. Interpreting solutions and deriving from those additional information. Developing differential-equations models for real-world phenomena.

Realism   Real-world application

Reactions   This is a fun, but nontrivial, project. Most students had little trouble with the first and second parts of the project, having seen parachute equations in labs and their textbook before using them here. The last part proved rather more difficult, and most students needed a poke in the right direction to come up with a way of making the force be continuous.

Credits   Inspired by ODE Models for the Parachute Problem, D. Meade, SIAM Rev. **40**(2):327–332

Technology   Numerical differential equations solver or computer algebra system.

> S.O.R. Tosane, Inc.
> 9999 N. 199th St.
> Lonlinc, SK 04685

Rigorous Mathematical Contractors,
Suite 3, Strawmarket Business Plaza
Lonlinc, SK 04685

Dear RiMaC:

As you may know, S.O.R. Tosane, Inc. is a newly founded company devoted to giving its well-paying clients the proverbial "ride of their lives," be it through scaling Everest or running through a crowded auditorium screaming "fire." A significant number of our customers are concerned with throwing themselves to the four-winds, so to speak—or at least, with throwing themselves out of planes, usually after having put on parachutes. In these endeavors we are as always most willing to help, though unfortunately it then becomes incumbent on us to investigate the safety of said clients, and it is in conjunction with this that we are contacting you.

Our well-heeled clients will be jumping from a military surplus cargo plane that we have purchased for this purpose. The plane will be flying at an altitude of around 10,000′, at a speed of approximately 175 mph, when they jump from the plane. All of our clients will be equipped with a main and reserve parachute; the main chute should be deployed early enough that if it fails they will be able to then deploy the reserve chute in time to avoid a messy end. We expect an average client to take on the order of 10 seconds to deploy the reserve chute if the main chute fails to deploy. The main chute takes around 3.5 seconds to fully deploy, while the reserve will deploy in only 1.5 seconds. Speed at impact should be about 15 feet/sec if the main chute deploys properly, or 17.5 feet/sec if the reserve chute is required. We expect the terminal (no pun intended) velocity in free-fall to be about 120 mph.

The first information that we need from you is when we should instruct our clients to deploy their main parachute so as to maintain a modicum of safety while catering to their daring bravado. In addition, we need to know where they will land, so that we know when to push them out of the back of the plane. It may be useful to note that your physical measurements department reports that air resistance on a falling body is proportional to the square of a person's velocity when no parachute is in use, and proportional to the velocity when falling with a parachute.

A secondary consideration is the stress that will be introduced on these eager clients in the course of their adventure. The "opening shock," or "jerk," (which, contrary to popular belief, refers not to the parachuter but to the effect to which they are subject—a subtle distinction lost on many), is defined to be the time derivative of the acceleration of the parachuter. We should like the force associated with the jerk ("jerk force") to be less than about 3 G/s over the time in which the parachute opens. (Here G is the force exerted by gravity on a body—thus for the force on the body to be less than 3 G, the acceleration must be less than 3 g.) You should therefore also indicate whether this is the case for our currently expected jump. We would also be interested in knowing if the "jerk force" is less than 1.5 G/s, so that we may allow older clients to be, if we may put it indelicately, jerked around.

Needless to say, the list of clients desiring to throw themselves from our plane is a long one. We therefore need your report on this matter in relatively short order. To facilitate this, we have arranged with a great friend and patron—this being none other than your peerless instructor—to answer any questions you might have in the process of working on this problem. We look forward to hearing from you.

Sincerely,
Sired M. N. Dehilary
President, S.O.R. Tosane Inc.

Solution

There are three parts to this project: first, finding the time at which the main parachute should be deployed, second, calculating the landing position, and third, determining the "jerk force." Allowing a discontinuous force on the skydiver, a common model for his or her vertical motion is the differential equation

$$v' = g - \begin{cases} k_1 v^2, & \text{before deployment,} \\ k_2 v, & \text{after deployment.} \end{cases}$$

k_1 and k_2 may be determined by using the terminal velocity before deployment (120 mph = 176 ft/sec) and assuming that the landing speed is the terminal velocity after deployment

(15 ft/sec or 17.5 ft/sec for the different parachutes). These give $k_1 = 0.001 \text{ ft}^{-1}$ and $k_2 = 2.15 \text{ s}^{-1}$ (for the main parachute) or 1.84 s^{-1} (for the reserve). It is then easiest to solve the differential equation numerically to determine the motion of the skydiver. To find the time at which the main parachute must be deployed, we find the distance that the skydiver falls if his or her main parachute fails to deploy and he/she then pulls his/her reserve, and require that terminal velocity for the reserve parachute be attained before impact with the ground. The distance fallen is the integral of the velocity (area under the graph of v against t). By numerically solving $v' = g - k_1 v^2$, $v(0) = 0$ (where $k_1 = 0.001$) we find the skydiver's velocity before deployment of a parachute. If we assume that this velocity has "reached" terminal velocity when it is within 1% of its final value, we find that this takes about 14.5 sec, in which time the skydiver has fallen approximately 1890 feet. If we assume that the skydiver realizes instantly that the main parachute has not deployed, he or she will take 10 sec to deploy the reserve and in that time will fall 1760 feet. Solving $v' = g - k_{2r} v$, $v(0) = 176$ (where $k_{2r} = 1.84$) similarly shows that the deceleration to within 1% of terminal velocity with the reserve parachute occurs in about 150 feet. Thus, of the 10,000 ft the skydiver falls, $1890 + 1760 + 150 = 3800$ ft is used in acceleration, parachute deployment, and deceleration. This leaves 6200 ft, which at 176 ft/sec takes 35 sec to fall. The latest the skydiver could wait before trying his or her main parachute is therefore $14.5 + 35 = 39.5 \approx 40$ sec. This is probably not the safest course of action, however.

The calculation of the skydiver's landing point may be done by assuming that there are no winds, that the skydiver does not make a concerted effort to change the direction of his or her travel, and that the drag model used above remains valid. Then the horizontal velocity of the skydiver is described by

$$v_h' = - \begin{cases} k_1 v_h^2, & \text{before deployment,} \\ k_2 v_h, & \text{after deployment.} \end{cases}$$

Solving this with a zero initial condition and then integrating to find distance travelled, we find that the skydiver goes about 2100 ft horizontally while falling—this is the distance at which horizontal velocity has gone essentially to zero, which takes about 30 sec.

Finally, to estimate the "jerk" we need to consider the time derivative of the acceleration of the skydiver—which is given by the right-hand side of the equation for his or her vertical velocity given above. If the force is represented by a step function this is infinite, which is problematic. To better model the force, we can use a continuous, smooth function to make the transition between the pre-deployment and post-deployment drag forces. This may be done by using time dependent drag coefficient k, so that

$$v' = g - \begin{cases} k_1 v^2, & \text{before deployment,} \\ k(t)v, & \text{during deployment,} \\ k_2 v, & \text{after deployment.} \end{cases}$$

We write the terminal velocity before deployment (176 ft/sec) as v_T, the time at which the parachute is deployed as t_D and the time for deployment as t_a ($= 3.5$ or 1.5 sec). Then the requirement that the drag be continuous and differentiable becomes the set of equations

$$k(t_D)v_T = k_1 v_T^2, \quad k'(t_D) = 0, \quad k(t_a + t_D) = k_2, \quad \text{and} \quad k'(t_a + t_D) = 0.$$

Assuming that $k(t)$ is a cubic in time (the lowest degree polynomial that can satisfy all of these conditions), we find

$$k(t) = -\frac{2(k_2 - k_1 v_T)}{t_a^3}(t - t_D)^3 + \frac{3(k_2 - k_1 v_T)}{t_a^2}(t - t_D)^2 + k_1 v_T.$$

Using this, we numerically solve for the vertical velocity and from that calculate the time-derivative of the acceleration to find the jerk. For the main parachute this has extreme values of about -80 ft/s^3 and 60 ft/s^3, or, given that $g = 32.2$ ft/s^2, about -2.5 g/s and 1.87 g/s. These thus turn out to be less than both the minimum (3 g/s) but not more stringent (1.5 g/s) criteria for the jerk that the skydiver should experience. For the reserve parachute the extreme values of the jerk are about -6.2 g/s and 4.3 g/s.

A few observations are worth making about this solution. It is clearly possible to choose the smoothing function to be

$$v' = g - \begin{cases} k_1 v^2, & \text{before deployment,} \\ k(t)v^2, & \text{during deployment,} \\ k_2 v, & \text{after deployment,} \end{cases}$$

but the algebra in this case becomes less tractable. It is also possible to model the drag function as being strictly linear, so that drag $= k_L(t)v$, where $k_L(t)$ is either a discontinuous or smooth function, as above. Finally, to be strictly accurate the calculations in the first and second parts should also be done using the modified equation derived in the third part, but we have not bothered to do that here, nor did students completing the project.

Fire Swamps and Rodents and Snakes (Oh, My)

Tommy Ratliff

Calculus I

Concepts   Modeling data with trigonometric functions

Realism   Completely fabricated

Reactions   The students needed some help in getting started on the functions to model the populations, especially to think of using trig functions rather than polynomials, but they generally did well with the mathematics after that point. They enjoyed the plot to this one, and the modeling really helped to reinforce some of the pre-calculus topics covered in the beginning of Calculus I.

Technology   Graphing calculator or computer algebra system

11 Patinkin Way
First National Park of Drachma
September 16, 1998

Math 101 Students
Wheaton College
Norton, MA 02766

Dear Calculus Students:

Things have finally quieted down around Drachma since the Prince was kicked out of office. The good news is that I've managed to find a government job as the head of the First National Park of Drachma. The bad news is that most of the Park consists of a Fire Swamp. When I went looking for help with our long range planning, your enterprising and resourceful professor naturally referred me to you.

We have two species that have me really worried about the future of the Park: the indigenous ELC (extremely large capybaras) and the brown tree snake which entered the Park about 50 years ago as a stowaway on a pirate ship. Fortunately, ELC's eat brown tree snakes. Unfortunately, brown tree snakes reproduce very rapidly.

My predecessor at the Park was a meticulous census taker (who used statistical sampling, by the way, to get more accurate counts), so I have approximate populations for each species for the last 30 years.

Year	Tree Snakes	ELC's
1968	15300	415
1970	9890	910
1972	2860	950
1974	3340	525
1976	9340	250
1978	12290	460
1980	9050	830
1982	4840	855
1984	5130	545
1986	8720	340
1988	10490	500
1990	8550	770
1992	6030	790
1994	6200	560
1996	8350	410
1998	9410	525

It looks like the populations are following some sort of pattern, but I'm not sure what it is. My real problem is that when either population gets very large, I will need additional employees to make sure that both species stay within the park and don't escape in the neighboring farmland. This is where I need your expert help (which your enterprising and resourceful professor assures me you can deliver). Specifically, I need a prediction for how large the populations will be in each of the next 20 years.

In addition, I believe the populations are fluctuating less and less, and may eventually stabilize. I would like your expert opinion on whether or not the populations do stabilize, and if they do, I need to know how long it will take and what the eventual populations will be.

Once the populations stop fluctuating so drastically, we will be able to dramatically improve access to the Park by offering summer camps, establishing permanent camp grounds, and perhaps even adding a logride, although there are still some flame-retardant issues to be worked out. This should all be possible when the ELC population is fluctuating by less than 75 per year and the brown tree snake population is fluctuating by less than 500 per year. As usual, I need your expert recommendation on when this will occur.

I have a meeting with the Budget Advisory Committee at the end of the month to propose our budget for the next two decades, so I would greatly appreciate your report by September 23.

Gratefully yours,
Amigo Flamboya

A Few Notes from Your Enterprising and Resourceful Professor

To see the general trend of the populations, I would suggest plotting the points for each population separately, with time on the horizontal axis and population on the vertical axis. It may make things a little bit easier if you let time $t = 0$ be 1968.

Solution

The answers to this problem will depend on what function the students use to model the data. The functions I used to create the points were:

For the ELC's: $g(t) = 450(.97^t) \cos\left(\frac{1}{5}\pi t - 2\right) + 600$.

For the tree snakes: $f(t) = 7500(.95^t) \cos\left(\frac{1}{5}\pi t\right) + 7800$.

Most of the students got something vaguely like this, and by plotting these functions, you can answer the questions:

Just plug into these equations to get the populations in the next 20 years (times $t = 30$ through 50).

The ELC population stabilizes to 600. The tree snake population stabilizes to 7800.

It will take roughly 40 years for the tree snakes to stabilize, and roughly 50 for the ELC population to stabilize.

Coffee to Go

Tommy Ratliff

Calculus I

Concepts   Newton's Law of Cooling

Realism   Completely fabricated

Reactions   Once the students set up the two-stage process, the calculations were fairly straight forward. This has been a popular project with the students. One group went so far as to contact a sibling in law school and discussed the legal liability of J.I.T. Box.

Technology   Scientific calculator

J.I.T. Box
132 Fast Food Lane
Partially Hydrogenated, WI 33021
November 3, 1997

Math 101 Students
Wheaton College
Norton, MA 02766

Dear Calculus Students:
After the verdict against one of our competitors a few years ago, my company, which owns a series of quick-dining establishments, has been hit with a series of copy-cat lawsuits. We need some technical expertise on one of these matters, and your enterprising and resourceful professor referred me to you.

The plaintiff, whom I will refer to as R. Clumsy for legal reasons, was the passenger in an automobile that stopped at one of our drive-through windows early one morning for coffee. Mr. Clumsy placed the coffee in a cup holder for several minutes, held the coffee cup in his hands for a short time, and then he spilled the coffee. He maintains that the coffee was much too hot, and he is suing for $200,000 for emotional distress and dry cleaning bills.

I would like to know how credible his story is. While we do serve our coffee at 160°F, I do not believe that the coffee was above the industry standard of 140°F when Mr. Clumsy spilled his coffee. He claims that it was exactly 7:58 AM when the coffee was poured, and when he spilled the coffee, some of it fell on his watch, a cheap Timex ripoff, which stopped at precisely 8:08 AM. He says that he left the coffee in the cup holder for 5 minutes, and held it in his hands for another 5 minutes before spilling it. I would like to be able to raise a reasonable doubt about the credibility of his story, either by showing that the temperature was *not* above the industry standard, or by showing that if the times are slightly different than he claims, then the coffee was not too hot.

Assuming Mr. Clumsy's scenario is correct, I would like to know if the temperature of the coffee was above the industry standard when he spilled it. I am doubtful that he remembered the times exactly. In particular, I seriously doubt that the coffee was in the cupholder and in his hands for *exactly* five minutes. What would the temperature be if the breakdown were 4 minutes in the cupholder and 6 minutes in his hands? 6 minutes in the cupholder and 4 minutes in his hands? 7 minutes in the cupholder and 3 minutes in his hands? I am also a little dubious about his claim that the coffee was poured at 7:58 AM exactly. What would the result be if the coffee were poured at 7:56 AM instead? In your report, I would like your expert opinion on whether or not there exists a reasonable doubt that the coffee was above 140°.

To help you out, the top investigators in our Fact Finding Department have uncovered the following information. Mr. Clumsy lives 7.8 miles away from the restaurant, so you can assume that the passenger compartment of the car had warmed to a comfortable 72°F, which is the same temperature that we keep the restaurant. Further, if you take a fresh, delicious, aromatic cup of our coffee and place it in a room at 72°F, then it will cool to 149°F in four minutes and thirty seconds. We also asked Mr. Clumsy to hold a thermometer in his hands just like he held the coffee. We discovered that the temperature was 92.3°F.

I understand that this is a busy time of the semester for you, but I would greatly appreciate your report by November 14, since the trial is set to begin the next week.

Yours sincerely,
J.I.T. Box

Solution

This is a two-stage process involving Newton's law of cooling: *The rate at which an object cools is proportional to the difference in temperature between the object and its environment.* In differential equations terms, we get $T' = k(T - T_e)$ with solution $T(t) = T_e + Ae^{kt}$ where k is the proportion, T is the temperature of the object, and T_e is the temperature of the environment. A key point is that the constant k is a physical property of the cup that will be the same for both stages.

In the first stage, you are given the initial temperature of the coffee and the temperature of the environment (72°). You can use the other information you are given to find the constant k, and then you are set to find the temperature of the coffee after it has been in the cup holder for five minutes.

Once he picks up the coffee, the second stage begins. The environment temperature changes to the temperature of his hands, and the initial temperature of the coffee is the temperature after 5 minutes in the cupholder. The results are

Minutes in cupholder	Minutes in hand	Temp when spilled (°F)
5	5	140.2
6	4	139.6
7	3	139.13
4	6	140.72
6	6	136.9

A Dorm Room's a Dorm Room, No Matter How Small

Tommy Ratliff

Calculus I

Concepts   Optimization of a function of one variable. The general case keeps students from using only graphing technology to find the minimum.

Realism   Completely fabricated

Reactions   The general case was challenging for many of the students.

Credits   This project was inspired by the Dome Tent project on page 104 of Cohen, Gaughan, Knoebel, Kurtz and Pengelley, *Student Research Projects in Calculus*, MAA, 1991.

Technology   A computer algebra system is helpful for the general case, but not strictly necessary.

Frieda Jo
Development Office
Dust-Mite University
November 2, 1998

Math 101 Students
Wheaton College
Norton, MA 02766

Dear Calculus Students:

After *months* of diligent work, I finally earned a promotion to Vice President for Development here at Dust-Mite U, which shocked quite a few because, as you may know, I am not more than two. I'm very proud of my fund-raising accomplishments, but sometimes the gifts come with very strict limitations on how they can be used. We just received such a donation, and when I went looking for help, your enterprising and resourceful professor naturally referred me to you.

We have a somewhat eccentric alum who has made a major contribution in memory of his favorite Chia Pet Airplane that recently passed away in a bizarre gardening accident (it's best we not discuss the details). As a fitting tribute to the dearly departed, the donor has designated that the funds be used to build a dorm in the shape of an airplane hangar, as shown below. There is an additional stipulation on the gift: the volume of the dorm must be *exactly* 225,000 cubic feet, which is one cubic foot for each sprout on the Chia plane.

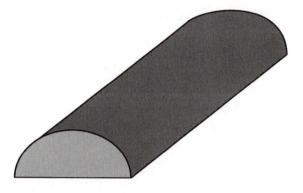

We're in the planning stages with the architects now, and we would obviously like to minimize the cost of the building. This is where I need your help. Currently, the construction costs for the foundation are $30 per square foot, the sides cost $20 per square foot to construct, and the roofing costs $15 per square. I need your expert advice on what the dimensions of the building should be to minimize the total cost.

While the cost of the flooring and siding has been fairly stable, a further complicating factor is that the cost of roofing material has been fluctuating dramatically for as long as I can remember (at least two months). In addition to your recommendation for the price of $15 per square foot, I also need a recommendation on the dimensions of the dorm if the roofing costs $R per square foot. We are meeting with the architects to discuss plans before Thanksgiving, so I would appreciate your report by November 11.

It's nap time now,
Frieda Jo

Solution

If you let r be the radius of the hemisphere and l the length of the building; then the cost equation for the specific case is

$$C(r) = \frac{1}{r}\left(\frac{27,000,000}{\pi} + 6,750,000\right) + 20\pi r^2,$$

which has a critical point at $r = 49.611$ and $l = 58.1978$.

In the general case, the cost function is

$$C(r) = \frac{1}{r}\left(\frac{27,000,000}{\pi} + 450,000R\right) + 20\pi r^2$$

which has a critical point at

$$r = \frac{1}{\pi}\sqrt[3]{5400\pi + 90R\pi^2} \quad \text{and} \quad l = \frac{450,000}{\pi r^2}.$$

The Broad Side of the Barn

Tommy Ratliff

Calculus I

Concepts   Riemann Sums

Realism   Completely fabricated

Reactions   The students had a fairly easy time understanding the questions that are asked. Most modeled the curved part of the barn as a semi-circle, although some did fit a parabola. Those who laid the siding in vertical strips generally had an easier time than those that used horizontal strips.

Technology   A graphing calculator is helpful, but not necessary.

Arnold Zinfandel
P.O. Box 2
Grand Island, NE 47318
December 2, 1998

Math 104 Students
Wheaton College
Norton, MA 02766

Dear Calculus Students:

I let my barn repairs go for too long, and now I find myself in a predicament. Although the barn is structurally sound and the roof is fine, the siding is rotting. When I went looking for some help in deciding on the best way to proceed, your enterprising and resourceful professor naturally referred me to you.

Mr. Chaney has offered to sell me aluminum siding that is either two and a half feet, one and a half feet, or one foot wide. I can order the siding in any length I need, and the cost is $3.25 per square foot. Since I can get Doug from over at the Edwards place to do the labor for next to nothing, my only real expense is the cost of the siding from Mr. Chaney. I want to do the entire barn in the same width siding, so for each width, I need to know what lengths of siding I should order and how much it will cost.

I know that there will be some waste material due to the shape of the barn. Being the environmentally aware individual that I am, I'd like to know how much siding will be wasted, and I'd also be interested in any suggestions you have for what I can do with the leftovers.

I've included a sketch of the barn below. Mr. Chaney has only guaranteed his prices through December 9, so I would greatly appreciate your report by then.

Yours sincerely,
Arnold Zinfandel

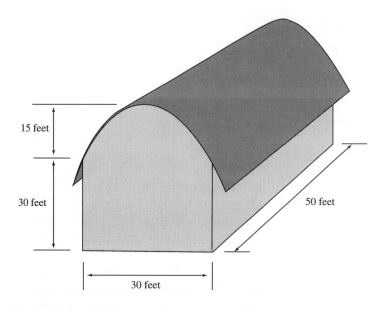

Solution

The only problem here is modeling the curved part of the barn. If you treat it as a semi-circle $y = \sqrt{225 - x^2}$, then you do a left sum to approximate $\int_0^{15} \sqrt{225 - x^2}\,dx$, and the different widths of siding determine the number of subdivisions. The final results are:

Siding Width	Amount of Siding	Waste	Total Cost
1 ft	5532 ft^2	25 ft^2	$23,512
2.5 ft	5563 ft^2	57 ft^2	$23,669
3 ft	5573 ft^2	66 ft^2	$23,686

A Question of Carpeting

Tommy Ratliff

Calculus II

Concepts   Approximating area using Riemann sums, error approximation

Realism   Completely fabricated

Reactions   Most students modeled the office as an ellipse (often after a reminder of the appropriate formula), although some did fit two parabolas together. The one-yard width caused some students problems.

Technology   A computer algebra system is helpful, but not necessary.

Galactic Gears, Inc
1 Galactic Way
City of Commerce, CA 55057
February 12, 1999

Math 104 Students
Wheaton College
Norton, MA 02766

Dear Calculus Students:

I recently became CEO of Galactic Gears after a somewhat messy hostile take-over, and I do not yet trust anyone at the company to help me with the critical decisions that will affect the success of the corporation. When I went looking for help, your enterprising and resourceful professor naturally referred me to you.

I am remodeling my (nearly) oval office and replacing the horrid orange carpeting left by my predecessor. With the Board and shareholders watching my every move, I want to be careful not to appear profligate. I have already decided on a tasteful aqua shag, but the carpeting comes on rolls in three different widths: 2.5 yards wide, 1.5 yards wide, and 1 yard wide.

I will use the same width for the entire office, but I'm not sure which one to choose. For each width, I need an analysis of how many square feet of carpeting I must buy, and an estimate of the amount of carpeting that will be trimmed and wasted. I also have the option to special-order the carpeting in any width. To impress the board with my thoroughness and all-around competence, I would like to know what width I should order so that no more than 30 square feet are wasted, and what width to order so that no more than 10 square feet are wasted.

I am including a sketch of my office, and I would appreciate an answer by February 23 since I need my office to be ready for a March 1 meeting with several venture capitalists to discuss the launching of the Gears Channel ("All Gears. All The Time"[TM]).

Yours sincerely,
M. Lugfield
CEO Galactic Gears

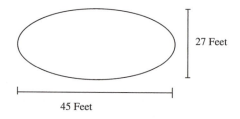

Solution

This project was assigned immediately after completing the chapter in Ostebee and Zorn, *Calculus from Graphical, Numerical, and Symbolic Points of View,* on determining the error bounds for Riemann sums, which was very useful in determining the widths necessary to waste less than 30 ft^2 and 10 ft^2. Measuring in feet, the shape of the office can be modeled by the ellipse

$$\frac{x^2}{22.5^2} + \frac{y^2}{13.5^2} = 1.$$

The exact area of the office is $(22.5)(13.5)\pi = 303.75\pi \approx 954.25$ ft^2. We can use a left sum on the integral $\int_0^{22.5} 13.5\sqrt{1 - x^2/22.5^2}\, dx$ to approximate the area of the upper right quarter of the room for the 7.5 and 4.5 foot widths. Since 22.5 is not evenly divisible by 3, we change the bounds to 1.5 to 22.5 and add the extra strip down the middle of the room. The totals are:

Width in Feet	Amount of Carpeting	Waste
7.5	1088.70 ft^2	134.45 ft^2
4.5	1044.00 ft^2	89.75 ft^2
3	1017.93 ft^2	63.68 ft^2

To find the widths so that no more than 30 ft^2 and 10 ft^2 are wasted, we can use the error bound formula given in Ostebee and Zorn (Theorem 1, Section 7.1, First Edition) that the left sum L_n approximates $I = \int_a^b f(x)\, dx$ within $|f(b) - f(a)|\frac{b-a}{n}$. Solving $|0 - 13.5|\frac{22.5-0}{n} < \frac{30}{4}$ gives $40.5 < n$. Then $n = 41$ and the desired width is approximately 6.5 inches. To waste less than 10 ft^2, the width needs to be approximately 2.2 inches. Notice that neither one of these is particularly practical.

The Great Escape
Tommy Ratliff

Course   Calculus II

Concepts   Volume of revolution

Realism   Completely fabricated

Reactions   The students needed a little prodding to develop the function to model the tank, but found the rest to be mostly straight-forward.

Credits   This project is essentially the same as the one on page 119 of Cohen, Gaughan, Knoebel, Kurtz and Pengelley, *Student Research Projects in Calculus*, MAA, 1991.

Technology   Graphing calculator

T. Houdini
135 Prestidigitator Avenue
Handcuff, WI 59055

October 15, 1997

Math 104 Students
Wheaton College
Norton, MA 02766

Dear Calculus Students:
I have decided to continue the family business established by my grandfather, and I need some help planning one of the escapes that I am including in my inaugural tour. When I went looking for help, your enterprising and resourceful professor naturally referred me to you.

I will be locked in chains and have my feet shackled to the top of a stool which is attached to the bottom of a giant tank that looks vaguely like a laboratory flask. The flask will be filled with water (at a constant rate of 500 gallons per minute), and after much practice out of the water, I have determined that it will take me exactly 10 minutes to escape from the chains.

I have a flair for the dramatic, so I would like to escape from the shackles at the exact instant that the water reaches the top of my head. I need your help in determining how tall the stool should be. Also, I want to monitor the rise of the water during the escape, so at any time after the water begins flowing, I want to know how high the water is in the tank and how fast the water is rising. While I am fairly accomplished at holding my breath under water, I would like to know how long I will have to hold my breath during the last part of the stunt.

I've included a sketch of the tank below, which gives the diameter of the tank at 1 foot intervals.

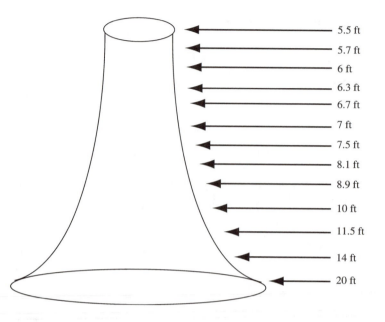

After consulting with your enterprising and resourceful professor, he suggested that you might be interested to know that I am 5 feet 9 inches tall, and I'm pretty skinny so that you can ignore both my volume and the volume of the stool in your analysis.

I realize that this is a busy time of year for you, but I would greatly appreciate an answer by October 24, since my tour opens at the Howard Johnson's in Kenosha on Halloween weekend.

Yours sincerely,
T. Houdini

A Few Comments From Your Enterprising and Resourceful Professor

After consulting with T. Houdini, I have a few suggestions that may help you get started:

- A gallon is equal to 0.13368 cubic feet.

- The tank looks very much like a solid formed by rotating a function about the y-axis. You will need to come up with a function to model this shape. You will probably want to keep this function as simple as possible.

- Next, express the volume of water in the tank as a function of the height of the water above ground level. What is the volume when the water reaches the top of Houdini's head? Once you have done this, you should be able to determine the height of the stool.

- You can think of the volume and the height of the water as functions of time. You can easily find an expression for $V(t)$, and then use your expression for volume in terms of height to solve for $h(t)$.

Solution

A function that fits the data fairly well is $f(y) = \frac{10}{\sqrt{y}}$ rotated about the y-axis where $1 \le y \le 13$. The volume at height h is given by $V(h) = \int_1^h \pi \frac{100}{y}\, dy = 100\pi \ln(h)$. After 10 minutes, there will be 668.4 cubic feet of water in the tank. Solving $V(h) = 668.4$ gives $h = 8.39$ so the stool should be 1.64 feet tall.

Notice that the volume after t minutes is $66.84t$. Setting this equal to $V(h)$ and solving for h gives $h(t) = e^{66.84t/100\pi}$. Since we are starting at $y = 1$, the actual height is $e^{66.84t/100\pi} - 1$.

I Guess Leonardo DiCaprio Was Booked

Tommy Ratliff

Calculus II

Concepts   Volume of revolution, surface area of revolution

Realism   Completely fabricated

Reactions   This was a challenging project for the students, especially since we had not discussed surface area of revolution beforehand. They also had some trouble finding the height of the water on the boat.

Technology   A computer algebra system is helpful for finding the height of the water on the boat.

Stanley
Eastland, CO 59055
March 12, 1999

Math 104 Students
Wheaton College
Norton, MA 02766

Dear Calculus Students:

Well, it's almost spring break here in Colorado—not that you'd ever know it, since it never stops snowing. Believe it or not, the whole class landed a guest shot on "The Love Boat: The Next Wave." It's gonna be really cool being on the cruise ship, but I do have some issues with the script. When I went looking for help, your enterprising and resourceful professor naturally referred me to you.

There is a lot about the script that I can't explain to you because I had to sign a standard non-disclosure agreement, but suffice it to say that the episode involves two of my class-mates I'll call "Jimmy" and "Parkman." The episode has a Titanic theme and Jimmy finds himself afloat in a concrete lifeboat in the North Atlantic. According to the script, Jimmy is doing fine until Parkman jumps from a flying saucer into the concrete raft and almost capsizes it.

I'm a little worried for Jimmy—after all, he hasn't had the best luck with stunts. I'm pretty sure that the boat is seaworthy (otherwise they wouldn't put Jimmy in that position, would they?) but I'd like your expert analysis to ensure that it will float. Since Parkman has added a little weight lately, I'm really concerned that he may actually capsize the boat. How much would he have to weigh for the boat to go under? Also, it isn't crucial, but I would feel much more at ease if you could let me know how close the water will be to the top of the raft before Parkman jumps in.

I've included a few 2-D sketches of the boat, and after consultation with your enterprising and resourceful professor, he suggested that you may need to know that the concrete hull weighs 40 grams for each square centimeter of surface area.

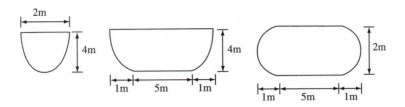

I know that you are about to start your spring break, but we leave to begin filming on March 30, so I would greatly appreciate your answer by then.

Yours sincerely,
Stanley

A Few Comments From Your Enterprising and Resourceful Professor

After consulting with Stan, I have a few suggestions that may help you get started:

- You will need to use Archimedes' Principal, which states that the buoyant force on an object in water is equal to the weight of the water displaced (*Calculus, Second Edition,* Hughes-Hallett, Gleason, et al., 1998, p. 383).

- The first thing to do is find the total surface area of the boat to determine how much it weighs. You will need to deal with the curved ends separately from the rest of the boat. Notice that the curved ends look a lot like a surface of revolution. You may need to look up the formula for the surface area of a surface of revolution in another calculus text. Ask me if you have questions about setting this up.

- Now you will need to find the volume of the boat to determine if the buoyant force of water is enough to support the boat. If you don't know this, a cubic meter of water weighs 1000 kilograms.

Solution

Notice that the ends of the boat look like the parabola $y = 4x^2$ rotated about the y-axis. The surface area of the center of the boat is

$$5 \cdot \text{Arclength} = 5 \int_{-1}^{1} \sqrt{1 + 64x^2} \, dx \approx 42.046 \text{ m}^2$$

The surface area of the ends is $\int_0^1 2\pi x \sqrt{1 + 64x^2} \, dx \approx 17.116 \text{ m}^2$. Since a square meter is 10,000 square centimeters, each square meter weighs 400 kg. Thus the total weight of the boat is 23,665 kg.

The volume of the center part of the boat is $(8 - \int_{-1}^{1} 4x^2 \, dx) \cdot 5 = 26.67$, and the volume of the ends is $\int_0^4 \pi \frac{y}{4} \, dy = 6.28$. Therefore, the total volume of the boat is 32.95 cubic meters. Therefore, the boat can support up to 32,950 kilograms. Therefore, the boat will float and Parkman would need to weigh 9285 kilograms in order to sink it. Not very likely, even for Parkman.

To find out how high the water will be on the boat, we need to find the height on the boat where the volume of the boat up to that height is exactly the volume of water displaced. Since the boat weighs 23,665 kg, it will displace 23.665 cubic meters. If we let the bounds on x go from $-t$ to t (rather than -1 to 1), then we can amend the calculations from the last paragraph to see that the total volume is $\frac{80}{3}t^3 + 2\pi t^4$. Setting this volume equal to 23.665 gives $t = 0.90123$. Therefore, the water will be 3.2488 meters high on the boat, or 0.7512 meters from the top.

While Visions of Emus Danced in His Head

Tommy Ratliff

Calculus II

Concepts   Infinite series, the harmonic series, the Integral Test for infinite series

Realism   Completely fabricated

Reactions   Students are very surprised to see infinite series pop up in this situation, and the result is quite non-intuitive. They find the project challenging, but have generally positive comments after completing the project. This is one of my favorite projects.

Credits   The idea for this project came from page 168 of Cohen, Gaughan, Knoebel, Kurtz and Pengelley, *Student Research Projects in Calculus*, MAA, 1991.

Technology   None required

Devious Dingo
Bleached Bones, Australia
April 27, 1999

Math 104 Students
Wheaton College
Norton, MA 02766

Dear Calculus Students:

HELP ME!! For the last two months, I've had this series of recurring nightmares that are about to drive me out of my mind! When I went looking for help, your enterprising and resourceful professor naturally referred me to you.

The scenario is nearly always the same. I'm standing at the end of a road that is 1 kilometer long (for some reason the road has those little green kilometer markers on it), and there at the other end is that @!*^@#! Emu, just standing there, sticking his tongue out. I start to go after him, but I can only run in slow motion, about 1 meter per second. After one second, *the road stretches **uniformly** and instantaneously by 1 kilometer* so now that pesky fowl is 1998 meters away, since some of the stretch happens behind me. I try to speed up, but I'm still moving in slow motion, at 1 meter per second. After another second, the road stretches again by 1 kilometer! And this just keeps on happening. Over, and over. And over. And over. Well, you get the idea. Then I wake up, hungry and frustrated.

I've gotta know: Do I ever get to the Emu? Do I have any chance? If I do get there, how long does it take? Should I take a snack to eat along the way?

Most of the dreams aren't that specific. Usually, I don't know how long the road is to begin with, or how fast I'm moving. All I know is that I'm always moving at the same slow rate, and the road stretches uniformly and instantaneously by its original amount after each second. You gotta help me figure out whether or not I get the silly bird, and if so, how long it will take.

I know it's the end of the semester, the weather is starting to get nice, and you're getting busy, but you've gotta give me an answer by May 7. I can't take this much longer.

Hungry as ever,
Devious Dingo

A Few Comments From Your Enterprising and Resourceful Professor

After reading Devious's sad tale, I have a couple of suggestions to help you get started.

- First, make sure you understand why the Emu is 1998 meters away after the first stretch.
- Next, set up a sequence $\{d_n\}$ where d_n represents the distance between Devious and the Emu after n seconds, but before the road does its instantaneous stretch. (e.g., $d_0 = 1000$, $d_1 = 999$, etc)

  Then write

  $$d_1 = 1 * \text{(some expression involving } d_0),$$
  $$d_2 = 2 * \text{(some expression involving } d_1),$$
  $$d_3 = 3 * \text{(some expression involving } d_2).$$

Now convert your expressions for d_2 and d_3 so that they only involve d_0. Use this to find a general expression for d_n in terms of d_0.

Solution

The students often have a hard time understanding why the Emu is 1998 meters away after one second. I usually show them how the stretch works by taking a rubber band and tying a paper clip near one end. As you stretch the rubber band, some of the stretch happens behind as well as in front. Then the students usually do some specific calculations for the first few stages and see that

$$\text{Position at time } n + 1 = \text{(Position at time } n) + \text{stretch} - \text{step}$$

or

$$d_{n+1} = d_n + \text{(\% of road in front)} * 1000 - 1$$
$$= d_n + \frac{d_n}{n * 1000} * 1000 - 1$$
$$= d_n + \frac{d_n}{n} - 1$$
$$= (n + 1) * \left(\frac{d_n}{n} - \frac{1}{n + 1} \right).$$

The first few cases give

$$d_1 = 1\left(d_0 - 1\right),$$

$$d_2 = 2\left(d_1 - \frac{1}{2}\right),$$

$$d_3 = 3\left(\frac{d_2}{2} - \frac{1}{3}\right).$$

Through substitution, we get $d_n = n(d_0 - \sum_{k=1}^{n} \frac{1}{k})$. Some may find it easier to express each d_n in terms of d_0 from the beginning, but this is how my students have usually approached the problem.

Since the harmonic series diverges, we will eventually get $\sum_{k=1}^{n} \frac{1}{k} \geq d_0$ so that $d_n \leq 0$ and the Dingo does catch the Emu. To see how long it takes, you do not want to attempt to find the partial sum where $\sum_{k=1}^{n} \frac{1}{k} \geq 1000$ (it will take a long time). Using the Integral Test, we know

$$\sum_{k=1}^{n} \frac{1}{k} > \int_1^{n+1} \frac{1}{x}\,dx.$$

Solving $\int_1^{n+1} \frac{1}{x}\,dx > 1000$ gives $n = e^{1000} - 1 \approx 1.970071114 \times 10^{434}$ *seconds* or $6.247054522 \times 10^{426}$ years. It's fun to have the students try to put this into a frame of reference, such as comparing it to the age of the earth.

The general case is very similar. If you let s represent the distance traveled each second, then $d_n = n(d_0 - s \sum_{k=1}^{n} \frac{1}{k})$ and Devious will catch the Emu in $e^{d_0/s} - 1$ seconds.

A Matter of Utmost Gravity

Tommy Ratliff

Multivariable Calculus

Concepts   Calculus of parametric equations

Realism   Completely fabricated

Reactions   It took the students a while to set the problem up, but then most were able to address the first case fairly easily. They found the general case more challenging, however.

Technology   A scientific calculator is required, although a computer algebra system is useful.

Devious Dingo
Bleached Bones, Australia
September 16, 1998

Math 236 Students
Wheaton College
Norton, MA 02766

Dear Multivariable Students:

Oh, the pain and agony. Day after day. Week after week. Year after year. I keep trying. But I just keep failing. After much experimentation, I've decided to focus on one method for gathering the main ingredient of my Aunt Edna's famous Emu Souffle: I'll use my handy catapult to hurl an anvil and squash the fowl as it comes around a blind curve. When I went looking for help in my endeavors, your enterprising and resourceful professor naturally referred me to you.

One of the complicating factors in cartoon land that you may not be aware of is that gravity not only has a vertical component, but also has a horizontal component. I'm a pretty smart carnivore, and I think could adapt my catapulting technique to this, except that GRAVITY KEEPS CHANGING! You know the way there are tide tables that tell you when high tide and low tide are? Well, here we have gravity tables that tell us what the horizontal and vertical components of gravity will be. There is *some* consistency in that gravity is always pulling down and toward the west.

So, here's my plan: I can set up the catapult near the blind curve due west of the location in the road where the Emu will be able to see me for the first time. Obviously, I'd like to squash the Emu at that spot in the road, and I'd like to be as far west as possible to increase the element of surprise. What I need to know is how far away I should set up the catapult, the angle that I should launch the anvil, and how long before the Emu reaches the spot that I should shoot it. Since I will launch it due east and gravity is pulling due west, I *really* want to avoid the classic anvil-goes-up-in-the-air-and-lands-on-poor-Devious-turning-him-into-an-accordion scenario. I'd like to know what angle I should ABSOLUTELY POSITIVELY avoid to escape this fate.

As regular as clockwork, the Emu zips down this road on Saturday mornings at 11:00 AM. I plan on launching my attack on September 26 (at 11:00 AM, of course) when gravity will be pulling down at a rate of 9.8 m/s^2 and to the west at a rate of 2.1 m/s^2. By the way, my trusty Acme catapult can launch an average-sized anvil at 45 meters per second.

I've been down the road enough times before (pun intended) to realize that everything may not go exactly as planned on the 26th. Since the gravity is not always the same at 11:00 on Saturday mornings, I would like formulas for each of my questions that depend only on the horizontal and vertical components of gravity. That way, I'll still be able to try my attack on another day.

In order to give me time to fine tune my instruments of destruction, I will need your report by September 25. Please don't let me down. Aunt Edna has been waiting a long time for this.

Hungry as ever,
Devious Dingo

Solution

If we assume that Devious is located at the origin, then the parametric equations for the path of the anvil can be found by using that the acceleration vector is $a(t) = (-2.1, -9.8)$, the initial velocity is $v(0) = (45 \cos(\theta), 45 \sin(\theta))$, and the initial position is $p(0) = (0, 0)$. Specifically the path of the anvil is given by

$$x(t) = -2.1\frac{t^2}{2} + 45\cos(\theta)t \quad \text{and} \quad y(t) = -9.8\frac{t^2}{2} + 45\sin(\theta)t$$

where θ is the angle of the anvil with the positive x-axis when it leaves the catapult. The anvil will hit the ground when $y(t) = 0$, giving $t = \frac{90 \sin(\theta)}{9.8}$. Substituting this value of t into $x(t)$ and differentiating gives that the maximum distance of 167.04 meters occurs when $\theta = .67985$ radians (about 38.9 degrees). Solving $x(t) = 0$ gives that 1.359 radians (about 77.8 degrees) is the angle where the anvil will land on top of Devious.

In the general case, the maximizing angle is

$$\theta = \arctan\left(\frac{-h + \sqrt{h^2 + v^2}}{v}\right)$$

(where h and v are the horizontal and vertical components of the gravitational acceleration, respectively) with a distance of

$$2025\frac{-h + \sqrt{h^2 + v^2}}{v^2}$$

meters. The angle to avoid is $\arctan(\frac{v}{h})$.

The *X-Files* Makes Me So Angry

Tommy Ratliff

Multivariable Calculus

Concepts   Optimization of a function of two variables

Realism   Completely fabricated

Reactions   The students really liked this project; they especially liked using the web to find the coordinates for each of the cities.

Technology   Internet access is needed to find the coordinates for each of the cities. A computer algebra system and the ability to graph contour plots is useful.

Roswell, NM 88201
October 21, 1998

Math 104 Students
Wheaton College
Norton, MA 02766

Dear Calculus Students:

I have been very upset with the one-sided portrayal of extraterrestrials in the popular media lately, so I have decided that it is time to present a kinder, more capitalist image. After studying the great hoopla here in Roswell recently, I have realized that the best way to destroy Earth is through retail. I received a grant from the Small Business Administration for legal aliens, which I plan to use to open several retail outlets. However, I need help determining where I should locate the central distribution site, and when I began looking for consultants, your enterprising and resourceful professor naturally referred me to you.

My main product will be a scaled-down version of the Intergalactic Detonator (more commonly used for removing view-blocking planets) which will be offered in select, premier outlets across the country. After careful and meticulous marketing research, I have decided to open stores in Nacogdoches, Texas, Mansfield, Massachusetts, and Kenosha, Wisconsin.

In order to keep inventory costs down, I want to have one central warehouse for the merchandise, and I will make weekly deliveries to each of the stores. I expect to fly three shipments per week to Kenosha, five per week to Mansfield, and four per week to Nacogdoches. Of course, I want to minimize the shipping costs, so this is where I need your help in determining where to build the distribution center so that the total flight distance is as small as possible.

If the market is as strong as I anticipate, I plan to open another outlet in Key West, Florida next summer. I expect the Intergalactic Detonator will be so wildly popular there that this store will require six weekly trips. Before I commit to opening a fourth outlet, I would like to know what effect this will have on the location of the warehouse. I need to know the optimal location for the warehouse if all four stores are open, and I would like your recommendation on which of the two warehouse locations is preferable.

In order to complete construction and be ready for the New Year's Eve rush, I would greatly appreciate your report by October 30.

Entrepreneurially yours,
Marvin

A Few Comments From Your Enterprising and Resourceful Professor

Here are a few suggestions that may help you get started:

- The first thing you will need to do is find appropriate coordinates for the cities in the xy-plane. You will need to find the distances between the cities, and then you can pick one of the cities to be located at the origin. After this, you can find the coordinates for the other cities.

- There are several web sites that may help you determine the distances between the cities. One that may be of particular use is The Great Circle Distance Calculator

$$\text{http://www.indo.com/distance/}$$

Solution

One set of coordinates that works is $(0, 0)$ for Nacogdoches, $(0, 848.255)$ for Kenosha, and $(734.86, 1286.38)$ for Mansfield. Then the function that represents the total distance traveled each week is

$$f(x, y) = 5\sqrt{(x - 734.865)^2 + (y - 1286.387)^2}$$
$$+ 4\sqrt{x^2 + y^2} + 3\sqrt{x^2 + (y - 848.255)^2}$$

which has a minimum at approximately $(179, 808.5)$. A very close city is Bluffton, Indiana.

The coordinates for Key West are $(924.549, -60.516)$ and the function to minimize becomes

$$f(x, y) = 5\sqrt{(x - 734.865)^2 + (y - 1286.387)^2} + 4\sqrt{x^2 + y^2}$$
$$+ 3\sqrt{x^2 + (y - 848.255)^2} + 6\sqrt{(x - 924.549)^2 + (y + 60.516)^2}$$

with a minimum at approximately $(500.5, 393.25)$. This corresponds closely to Jackson, Georgia.

The Company Picnic

Elyn Rykken

Survey of Mathematics Course

Concepts   Using set theory and Venn diagrams to organize and manipulate data.

Realism   While this project is completely fabricated, it strikes me as a plausible use for Venn Diagrams.

Reactions   This is the first project that I assigned in the course. As is the case with all of the assignments written for a "survey of mathematics" course, the mathematics is not complicated. While most students obtained the correct answer, writing up their explanation was not always easy for them.

Technology   None required

International Enterprises
123 Lower Wacker Drive
Chicago, IL 60601
June 25, 1998

Mathematics Department
Your University
100 Main Street
College Town, USA 45678

Dear Math Students:

I recently landed a job at International Enterprises. Part of my job responsibilities include organizing the company picnic for the 4th of July. I have run into some trouble with this. I called the math department at your school and they referred me to your professor who referred me to you.

Part of organizing the picnic includes ordering the food. To determine what people wanted to eat at the picnic, I took a survey of the 256 employees. I asked them if they wanted a hamburger or a hot dog. After collecting the surveys, I had my secretary add up the number of hamburgers and the number of hot dogs that had been requested. 197 people had checked the hamburger box and 182 had checked the hot dog box. 33 people phoned me and told me that they were vegetarian and they wanted garden burgers instead. I thought that everything was fine until my new boss told me that due to budget cuts, people were only allowed either a hamburger or a hot dog, but not both. I asked my secretary if he still had the surveys, but he told me that he had recycled them. When I phoned my boss, she also told me that it would be too expensive to start over and send out another survey. Can you help me?

I figure that those who ordered both won't really care if they get a hot dog or a hamburger, but I have no idea how to figure out who ordered both. Given this, could you let me know how many people ordered both? Could you also list some of the possible combinations of numbers of hot dogs and hamburgers that I could order?

Since you are college students, I have every faith in your abilities to help me out. Since the picnic is Friday, July 5th and I need to place my orders by Wednesday, July 3rd, could you please get back to me by Tuesday, the 2nd? Thanks for all of your help.

Yours sincerely,
J.R. Doe
International Enterprises

Solution

156 employees checked both the hamburger and the hot dog boxes. 41 employees wanted only a hamburger, and twenty-six wanted only a hot dog. There can be between 41 and 197 hamburgers and between 26 and 182 hot dogs as long as their total is 223. (Thirty-three veggie burgers should be ordered no matter what.)

The Restaurant

Elyn Rykken

Course   Survey of Mathematics Course

Concepts   Using expected value and other basic concepts of probability.

Realism   Completely fabricated

Reactions   The students generally organized the material nicely using tables.

Technology   Simple calculator

Pizzas Plus
100 Michigan Avenue
Chicago, IL 60601
July 9, 1998

Mathematics Department
Your University
100 Main Street
College Town, USA 45678

Dear Math Students:

A friend of mine, J.R. Doe, told me how you helped her solve a tricky problem concerning her company's fall picnic. The picnic was a big success and J.R. has been promoted. I also have a problem that I was hoping you could help me solve.

I recently bought a restaurant in Chicago. It is pretty small, but well-located. We are open every day, but Mondays, from 4 PM until 11 PM. Our menu is basically Italian. We offer individual pizzas, pastas, salads, and a special of the day. We also have really good desserts. We offer cookies, ice cream and superb cakes. Last month we had 800 customers. Our salads are considered a whole meal, so people usually don't order another entree with them. Last month, we sold 400 pizzas, 200 pasta dishes, 130 salads and 70 specials of the day. Not everyone orders a dessert, but last month we sold 200 cookies, 100 ice creams and 250 cakes. Next week, we are expecting 250 to 300 customers. I need to know how many of each of the entrees and how many of each of the desserts I can expect to sell.

I would also like to know how much money I can expect to make. Perhaps you could also let me know how much the average customer spends. Let me tell you our prices. The pizzas are $7.95. The pastas are sold for $8.95. The salads are $6.95 and the special of the day is $9.95. As for the desserts, the cookies are $1.00, the ice cream is $1.50 and the cakes are $3.00. Thanks a lot for your help. I am not very good at math, but perhaps you could try to explain your answers in such a way that I could estimate them for myself next time.

Yours sincerely,
Chris Smith
Pizzas Plus

Solution

Chris should expect to sell 125 to 150 pizzas, 63 to 75 pastas, 41 to 49 salads, and 22 to 26 specials. He should also expect to sell 63 to 75 cookies, 31 to 38 ice creams, and 78 to 94 cakes. The average customer spends $9.59 so Chris can expect to make between $2,397.50 and $2877 next week.

The Commute

Elyn Rykken

Course   Survey of Mathematics Course

Concepts   Manipulating basic formulas and unit conversions

Realism   Real-world application with real data

Reactions   While I did not drive a white Mazda, at the time I wrote this, this project described my actual commute. In addition to expressing their surprise at this, my students often suggested that Professor Johnson should buy the VCR in a state with lower sales tax, such as Indiana.

Technology   Simple calculator

Indiana University Northwest
3400 Broadway
Gary, IN 46408

August 1, 1998

Mathematics Department
Your University
100 Main Street
College Town, USA 45678

Dear Math Students:

A friend of mine, Chris Smith, told me how you helped him solve a tricky problem concerning his new restaurant, Pizzas Plus. I also have a problem that I was hoping you could help me solve.

I recently finished my Ph.D. in history from the University of Chicago. This fall, I got a job teaching at IUN. I still, however, live in Illinois. My three classes meet on Tuesdays and Thursdays. I commute to school each Tuesday and Thursday using Highway 94. The drive is 51 miles each way and takes me between an hour and an hour and fifteen minutes depending upon traffic. Since I live so far away from school, I don't come in on the days I don't teach. I need to know how much money I should budget for gas and oil changes for work for the fall semester.

Let me give you a few more details. My car is a white, 1991 Mazda which gets around 30 miles per gallon. I buy my gas from the Shell station close to school. The average price for gas has been $1.31 a gallon. I get my oil changed at the Lube Pro's near my house and they charge $24.95 including taxes for an oil change and check-up. I change my oil roughly every 3,000 miles. The semester is 16 weeks long including finals, and we have Thanksgiving day off.

I hope that I have given you all the information that you need. I was hoping that you could respond by August 15th. I have $400 left in my budget to cover commuting expenses and I need to know if I'll have enough money left over in my budget to buy a new VCR. The VCR I want to buy is $219 plus the 8.5% sales tax. It would be really helpful if you could figure out whether or not I should buy the VCR as well. Thank you very much for all of your help. I know that you are near the end of a semester and must have a lot of homework.

Yours sincerely,
Professor Johnson
IUN

Solution

I often changed the prices for gas and the VCR depending on the current prices at the time of the assignment. In this version, Professor Johnson should expect to spend $163.02 on commuting. This leaves $237 dollars for the VCR, which in Illinois would cost $237.62. Most of my students said that Professor Johnson could afford the VCR, especially if he or she bought it in Indiana.

The Raise

Elyn Rykken

Survey of Mathematics Course

Concepts   Computing annuities

Realism   Real-world application with fabricated data

Reactions   Most students felt that this was a valuable project with applications to their own life.

Technology   Scientific calculator

International Enterprises
123 Lower Wacker Drive
Chicago, IL 60601
August 8, 1998

Mathematics Department
Your University
100 Main Street
College Town, USA 45678

Dear Math Students:

Professor Johnson asked me to tell you that he is very pleased with the help that you gave him on his budget. He is now enjoying his new VCR. Partly because of the success of the picnic, I was given a promotion and a raise. Now that my future with the company seems secure, I would like to buy a house and perhaps a car. I was hoping that you would be able to help me with my finances.

With my raise, I will be earning $30,000 a year. I called a bank and they told me that the maximum amount of money that I can borrow from a mortgage company is one whose payment is 28% of my monthly income. Could you figure out how much I could afford for a monthly payment? Currently, I am renting a one bedroom apartment for $600 a month. The bank that I phoned had 30-year mortgages at 8% per year. If I get a mortgage from them, how much would I be able to borrow to buy a house?

Unlike Professor Johnson, who drives a 1991 Mazda, I drive a 1985 Toyota and am in need of a new vehicle. I have been looking at the new Mazda Proteges. They are selling for about $16,000 and are currently being offered with 5-year loans with monthly payments at 3% per year. If I bought the car with these terms, what would my monthly payments be? I think that I could afford as much as $300 a month.

Thanks again for all of your help. I don't know what I'll do when the semester is over.

Yours sincerely,
J.R. Doe
International Enterprises

Solution

J.R. Doe can afford $700 a month for his or her mortgage payment. Using the formula

$$P = R \left[\frac{1 - \frac{1}{(1+r)^n}}{r} \right]$$

where P is the present value of the annuity, R is the monthly payment, n is the number of payments, and r is the interest rate per period, we have $R = \$700$, $n = 360$ and $r = \frac{.08}{12} = .0066667$. This gives us $P = \$95,398.45$ as the amount J.R. Doe can borrow to buy a house. To calculate the amount needed to purchase a \$16,000 car we use the formula

$$R = \frac{P}{\left[\frac{1 - \frac{1}{(1+r)^n}}{r} \right]}$$

where $P = \$16,000$, $r = .0025$ and $n = 60$. This gives us a monthly payment of \$287.50.

The Car

Elyn Rykken

Survey of Mathematics Course

Concepts   Elementary modeling using linear and exponential functions.

Realism   Real-world application with real data

Reactions   This project was difficult for most of my students, especially when it came to having to offer another reasonable way to predict the car's value.

Technology   Simple calculator

International Enterprises
123 Lower Wacker Drive
Chicago, IL 60601
August 8, 1998

Mathematics Department
Your University
100 Main Street
College Town, USA 45678

Dear Math Students:

Professor Johnson asked me to tell you that he is very pleased with the help that you gave him on his budget. As for me, partly because of the success of the picnic, I was given a promotion and a raise.

I recently traded in my 1985 Toyota for a 1997 Volkswagen Jetta. I was hoping that you would be able to help me with my finances by helping me plan for the future. I am interested in predicting how much my Volkswagen will be worth in the upcoming years and have no idea how to go about getting such a prediction.

A new 1997 Jetta sold for approximately $19,000. This year, a used one is worth roughly $15,000. The car was and still is in excellent condition. I always maintain my cars. That means that it lost $4000 in value last year. I am not sure if it will continue to depreciate at this rate. If it does, can you give me a formula to calculate the car's value after a certain number of years? Another way to consider the problem is that the car is now worth roughly 78.9% of its original value. Perhaps next year it will only be worth 78.9% of its current value. If this is the trend, can you give me a different formula to calculate its value? Which of these formulas do you think will better predict the value of the car? Perhaps you could compare the values you get for different years. Do you know of any other ways to obtain such a prediction?

Thanks again for all of your help. I don't know what I'll do when the semester is over.

Yours sincerely,
J.R. Doe
International Enterprises

Solution

Let's consider $L(t)$ as $t = 0$, where t is the number of years J.R. has owned her car. The linear approximation for the value of the car is $L(t) = 15,000 - 4,000t$. Using this model, the car will be worth nothing in only 4 years. The exponential model is given by $E(x) = 15,000(.789)^t$. This is certainly a more realistic long-term model.

The Square Meal Deal

Elyn Rykken

Survey of Mathematics Course

Concepts   Understanding the multiplication principle and the difference between permutations and combinations.

Realism   Real-world application with fabricated data

Reactions   This was a difficult assignment for my students to solve and to explain. Many did not calculate the number of choices correctly.

Credits   Thanks to Iztok Hozo of Indiana University Northwest for giving me the reference.

Technology   Simple calculator

International Enterprises
123 Hamilton St.
Allentown, PA 18104
June 25, 1998

Mathematics Department
Your University
100 Main Street
College Town, USA 45678

Dear Math Students:

I recently landed a job at International Enterprises. Part of my job responsibilities include organizing an advertising campaign for one of our clients. I have run into some trouble with this. I called the math department at your school and they referred me to your professor who referred me to you.

The restaurant chain *Chicken Licken* has asked our company to write and design their new ads. They offer a deal where for $8.95 the customer chooses one of five main courses: a chicken sandwich, a chicken breast, a chicken pot pie, chicken sticks, or vegetarian lasagna. With each meal the customer also has the choice of two side dishes. The restaurant offers French fries, mashed potatoes with gravy, baked potatoes, hash browns, bread, salad, soup, biscuits, cottage cheese, or fruit. So in total, there are ten side dishes. The meal also comes with dessert. Here there are four choices: pie, ice cream, cake, or cookies. As part of the ad, the company wanted to know the number of different possible meals. After consulting a book, I calculated that there were 1800 possible meals, which seemed like a nice number to use in advertising.

I told a friend about my project and she said that she had read an article ("Teacher's Diligence Finds Fame, Free Lunch," *The Morning Call* (Allentown), Jan. 21, 1995, Joseph P.

Ferry) about how Boston Chicken (now known as Boston Market) had run a similar ad and had miscalculated the number of possible meals. Apparently they offered sixteen side dishes and the customer was allowed to choose three of them with each meal. The company's ad (apparently featuring quarterback Joe Montana) claimed that there were 3,360 possible combinations of three side dishes chosen from the sixteen offered. A high school mathematics teacher, Bob Swaim, convinced them that they had made a mistake. He and thirty of his students received a free lunch from Boston Chicken. He also appeared on the CBS program "Good Morning America." The teacher argued that whoever had done the mathematics had confused the concepts of permutations and combinations. By his calculations there were only 816 different combinations available. He also noted that the ad did not allow for the possibility of choosing a side dish more than once and that his formula did. After hearing about this I began to question the calculations that I had made. I don't want to be embarrassed by having the same thing happen to Chicken Licken.

Since you are college students, I have every faith in your abilities to help me out. I need to know many possible meals there are. I would also really appreciate it if you could tell me how Boston Chicken and Bob Swaim arrived at their different answers. Thanks for all of your help.

Yours sincerely,
J.R. Doe
International Enterprises

Solution

J.R. Doe has made the same mistake that Boston Chicken did. Boston Chicken used permutations and calculated that the number of possibilities were $16 \times 15 \times 14 = 3360$. They should have used combinations. If they allow for the possibility of choosing an entree more than once, the answer would be $\binom{16}{3} + 2\binom{16}{2} + \binom{16}{1} = 560 + 240 + 16 = 816$. In this same spirit, J.R. Doe calculated the number of possibilities using $5 \times 10 \times 9 \times 4 = 1800$ when he or she should have written $5 \times \left[\binom{10}{2} + \binom{10}{1}\right] \times 4 = 1100$.

Stuck in Traffic in Chicago

Elyn Rykken

Course   Calculus II

Concepts   Uses Riemann Sums

Realism   Fabricated application with real data

Reactions   While the travel time the students calculate is only an estimate, some students correctly point out that they are only given the speeds at one particular moment in time and not at several different times and hence, their estimate is not as accurate as it could be. More commonly, students make the mistake of incorrectly calculating an "average" speed by adding up the speeds and dividing by 21, then dividing the distance by this rate to calculate the travel time.

Credits   This project was written jointly with Maureen Carroll, *University of Scranton*. The data was taken directly from the Illinois Department of Transportation Traffic Systems Center (TSC) at their web site `http://www.ai.eecs.uic.edu/GCM/dan-ryan.html`.

Technology   Nothing more than a simple calculator is needed, although a spread sheet program such as Excel is very useful.

Barney, Smith and Elmo
123 Lower Wacker Drive
Chicago, IL 60601

February 2, 2000

Mathematics Department
Your University
100 Main Street
College Town, USA 45678

Dear Calculus students:

I am a lawyer for the firm of Barney, Smith and Elmo. One of my clients has been accused of robbing the downtown Chicago branch of MegaCityBank on January 27th of 2000. My client claims that he left work and went directly home to study for an exam in his evening class. The police claim that he robbed the bank on his way home. Allegedly, my client bears some resemblance to the thief caught in the act by the bank cameras. I need to convince a jury that my client did not commit this crime, but I can't do this without your help. I have gathered a lot of evidence concerning the events of the 27th, but I don't know how to piece it all together. I haven't had a mathematics course in years and, although I'm embarrassed to admit that I can't solve this puzzle, I'm willing to ask for help.

Allow me to explain the events of January 27th. A validated parking ticket with an exit time of 1:44 PM stamped on it shows when my client left the parking garage in his building. The bank robbery took place at exactly 2:02 PM as shown by the video cameras at the bank. Since the parking garage clock runs a minute behind the bank clock, my client had 17 minutes to get from his building to the bank. MegaCityBank is located just off of the Dan Ryan Expressway at Roosevelt Road, right in the heart of downtown Chicago (which is what Chicagoans call "the Loop.") My client works south of the Loop, and the Dan Ryan Expressway is the quickest route from his workplace to the Loop. It takes my client about one minute to enter the Dan Ryan Expressway at 95th Street and head northbound towards the Loop, where MegaCityBank is located. As part of my client's defense, I want to show that travel times on the Dan Ryan were such that it would have been impossible for him to commit the crime. Right now you're probably thinking that securing this type of information is nearly impossible. Fortunately, this isn't the case. The World Wide Web and the Department of Transportation (IDOT) have come to our rescue.

The Gary-Chicago-Milwaukee Corridor Transportation Information Center provides expressway traffic information. In fact, it gives travel times to and from the Loop for all of the major expressways in Chicago. The Center's computer receives raw data from the IDOT Traffic Systems Center (TSC) once every minute. The TSC receives its data from loop detectors embedded in the pavement on the expressways. The loop detectors act like metal detectors and can sense when a vehicle passes over them. This allows the TSC to approximate the speed of traffic at each loop detector. Simple formulas have been developed to convert this data into travel time and congestion estimates. A more extensive explanation for how the system works can be found at their web site at `http://www.ai.eecs.uic.edu/GCM/GCM.html`.

I have included a copy of the data from the Dan Ryan Expressway Loop Detectors from January 27th at 1:45 PM. I am interested in determining the northbound travel time from 95th Street to Roosevelt Road using the data from the detectors. Now, to complicate matters, the Dan Ryan has both express lanes and local lanes and the travel times for these are typically not the same. Heading northbound the expressway splits into express and local lanes at 65th Street. So, after entering the Dan Ryan at 95th Street my client could have used either the local or the express lanes starting at 65th Street. Also, travelers can only exit at Roosevelt Road from the express lanes. So, if he did use the local lanes then he would have to switch back to the express lanes by 29th Street in order to be able to exit at Roosevelt Road. So, I am also interested in determining the travel time if my client took the northbound local lanes from 65th Street to 29th Street and used the expressway lanes for the rest of the trip. I don't know if this will be helpful, but I drove the Dan Ryan from 95th Street to Roosevelt Road and the distance is approximately 10.4 miles. The express and local lanes are just separated by a concrete median, so the distance is the same for either.

As I told you above, although I took calculus in college many years ago, my knowledge of mathematics is limited. Please include a clear and thorough explanation of how you arrived at your answer. Be sure to include and explain any formulas that you may have used. I have to understand your solution completely in order to explain it to a jury. Thanks for your help.

Yours sincerely,
G. E. Toffelhook, Esquire

Data for Northbound Dan Ryan Expressway Loop Detectors
on January 27th, 2000 at 1:45 PM

NB Dan Ryan Expressway		NB Dan Ryan Local Lanes	
Street	mph	Street	mph
95th Street	34		
90th Street	29		
87th Street	23		
83rd Street	23		
79th Street	32		
75th Street	9		
71st Street	37		
65th Street	55	65th Street	14
59th Street	55	59th Street	55
55th Street	55	55th Street	19
51st Street	55	51st Street	23
45th Street	55	45th Street	19
Root Street	55	Root Street	21
35th Street	30	35th Street	55
33rd Street	55	33rd Street	13
29th Street	48	29th Street	15
26th Street	16		
24th Street	21		
21st Street	10		
16th Street	16		
Roosevelt Road	28		

Solution

The table of sample data is for the twenty-one detectors between 95th Street and Roosevelt Road. This allows students to use twenty intervals in their calculations. To simplify calculations, we suggest that students assume the detectors are more or less evenly spaced. If one wishes to update the data, the northbound Dan Ryan is most congested in the mornings and early afternoons. For the current set of data, using either the left-hand rule, right-hand rule, trapezoid rule or Simpson's rule, the travel time is approximately 24 minutes. For example, to obtain the left-hand approximation, we multiply the reciprocal of the velocity at the beginning of each interval by .52 miles and then sum.

3

Sample Solutions

In this chapter we include two sample solution papers, with comments on how they might be graded, one using a checklist and one using a rubric. In the first case, this is an actual solution paper that one of the authors received, while the other is a fabricated paper that illustrates the type of project solutions that we are accustomed to receiving. In both cases the reader will note the changes that would change an adequate solution paper into a very good one. We have found that the papers we receive are, on the whole, of quite good quality—especially as students become familiar with this type of assignment, the fact that they put significant work into their final solution results in the average paper being quite good.

3.1 Solution Paper and Checklist —Annalisa Crannell

When I first started grading papers, I did it somewhat differently than I do now. I used to read through the papers much more thoroughly. This allowed me to change my checklist over the years to make the questions more consistent with what I actually wanted.

The sample student paper following is for the project "The Case of the Crushed Clown," which is from my early years of grading projects—in fact, it is the third paper I assigned during my first semester at Franklin & Marshall College. In the description below, I explain how I would approach this paper if I received it in a class today. I would skim through it as described, and sometimes would go on to read every word carefully. What has not changed is that I try to write almost all of my comments on the checklist, and to include very few comments on the paper itself.

In the subsections that follow are the sample solution paper, the checklist used to grade it, and comments regarding the grading.

3.1.1 Solution Paper

Todd Bartos
Franklin and Marshall
Math 110
11/24/92

Officer S. Kovalevskia
c/o Police Department
Constabulary Avenue
Big City, PU 11235

Dear Officer Kovalevskia:

I am very sad to hear of the demise of Bobo the clown. I knew Bobo personally, and I can tell you that he did not kill himself. He is not the crying-on-the-inside kind of clown, but rather loved life and lived it to the fullest, even if on the edge. In fact, I believe that Bobo was killed, and I will tell you how.

In your initial letter to me, you told me that you had discovered that the cannon shot Bobo out at 30 meters per second (m/s), a very considerable speed. I was also told that the cannon is three meters off the ground, and from base to tip, the cannon is five meters long; which means that the cannon covers four meters of horizontal distance (see figure III). I understand that poor Bobo hit the canyon wall opposite the cannon 310 meters down from the cannon base. As near as I can tell, the crime looked something like:

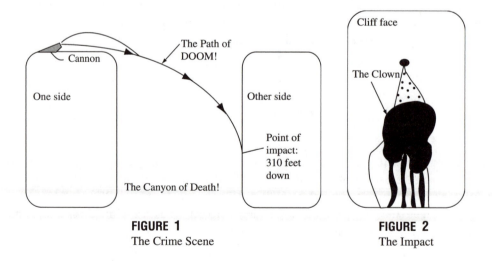

FIGURE 1
The Crime Scene

FIGURE 2
The Impact

I really feel for Bobo's fans, as their hero is now what appears in figure II. Now, as for Bobo and his journey.

Bobo left the cannon at 30 m/s. If this is treated as a vector, or a direction with a value (in the figures, vectors are represented by arrows), then the cannon can be used as a "vector" triangle as below:

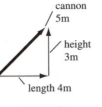

FIGURE 3
The Cannon

This represents the 3, 4, 5 Pythagorean triple. The triple means that the sum of the squares of the smallest and next greatest sides is equal to the square of the hypotenuse, or the long side. Simply stated, $3^2 + 4^2 = 5^2$. This triangle is the basis for all calculations related to Bobo's murder. If the vectors are changed to reflect velocity, then the diagram becomes:

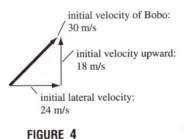

FIGURE 4
The Conversion to Velocity

Each vector has been multiplied by a factor of six, as $30/5 = 6$. This division is done because I wanted to find a vector triangle with the same relationships as the 3, 4, 5 triple. The 30 m/s represents the initial velocity, or $v(0)$. The 18 m/s represents the initial velocity in the vertical, or "y" direction, or $y'(0)$. The 24 m/s represents the initial velocity in the horizontal direction, or $x'(0)$. The prime notation is used because velocity is the derivative of position. Bobo was at the base of the cannon $(0, 0)$ when fired, so the velocities in the diagram are for when time equaled zero. In order to find how fast Bobo was going, I needed to find the initial values of the cannon.

The second derivative of the position, or the first derivative of the velocity is the acceleration, or $v'(t)$, $x''(t)$, and $y''(t)$ respectively. Acceleration due to gravity is a constant at 9.81 m/s. Since gravity is a force that "pulls down," the actual value for this is -9.81 m/s. This occurs in the "y" direction, so therefore $y''(t) = -9.81$ m/s. Because of the unique nature of Bobo's flight, there was no linear acceleration throughout the flight. If you need proof of this, take a sockball and toss it to someone. While it appears to slow, it is not true. The rate at which the sock travels to the other person is constant, only the vertical motion is changed, and that is due to gravity. Therefore, $x''(t) = 0$. To find the first derivative of position, I integrated the accelerations. When integrating an indefinite integral, a constant must be added at the end in order to compensate for the interval. For x, the constant is 24, and for y, the constant is 18. This comes from then initial velocities, i.e., at time zero. By integrating the second derivative, $x'(t) = 24$; and $y'(t) = 9.81t + 18$. By integrating these equations again, the position can be found. The constants for each equation are zero, but only the $y'(t)$ need be integrated, as I needed to find the time when Bobo hit the wall, 310 m down. The equation for position is: $y(t) - 4.9t^2 + 18t$. By setting this equation equal to -310 m, It is possible to solve for time.

$$-310 = -4.9t^2 + 18t$$

$$0 = -4.9t^2 + 18t + 310$$

$$t = -6.32, 10.$$

Since time can't have a negative value in this case, then Bobo hits the wall ten seconds after he is shot out of the cannon.

To find the velocity at which he hit the wall, plug 10 in for t in each of the velocity equations, find the x and y for which $t = 10$, and the square both values and add them together, then take the square root of the new number. This is done to find the velocity vector (see figure IV) given x and y.

$$x'(10) = 24 \text{ m/s} \qquad y'(10) = 81.9 \text{ m/s}$$

$$x'(10)^2 = 576 \text{ m}^2/\text{s}^2 \qquad y'(10)^2 = 6707.6 \text{ m}^2/\text{s}^2$$

$$\sqrt{576 + 628373.3} = 85.3 \text{ m/s.}$$

Thus Bobo hit the wall going approximately 85 m/s, which is, to say the very least, exceptionally fast. Therefore, Bobo died from contact with the wall, not before entering the cannon.

And now the killer ... (suspense). Rick Rasterdly was indeed filming the elephants, but he deliberately set the video camera's timer to be slow. What he did not count on was the fact that Bobo's watch was dead-on exact (no pun intended). The video camera read 15:52:20 at the time of the accident. If you are correct about the time difference, then the camera should have read 15:56:36. When Bobo was scraped off the canyon wall, his watch read 3:56:50. From what I have already found out, Bobo took flight at 3:56:40, as it took ten seconds to fly into the wall. This means that Mr. Rasterdly had three and one-half seconds to drop the camera and get to the button, since it takes one-half of a second to push the button. There is your killer, Officer Kovalevskia. Just to add a little drama to the trial, you will be able to present the fact that Bobo flew 240 m across the canyon. If you do your research, you will find that this breaks the world record for flight by a man from a cannon. I was able to find this by integrating the x component of the velocity vector, so $x(t) = 24t$ and putting in ten for time. The constant drops out because $x(0) = 0$, and the linear distance is then 240 m.

I hope that I have been of service to you and the Police Department. I want to see justice done and Rasterdly to the slammer. If there is anything I can do for the trial, please do not hesitate to let me know.

Sincerely yours,
Todd Bartos

3.1.2 Grading and Comments

The checklist that I hand out, which is given in Table 1, asks students to staple their papers, to number the pages, and to attach the checklist to the top of their paper with a paperclip. Todd did not number his pages, so I would use red pen to circle the words "number your pages" on the checklist. (I separately circle any directions that a student did not follow.)

Now I am ready to begin grading. What I try to do is to look at the questions on the checklist and find out whether the answer is **YES** or **NO**. I will count up all the **YES**'s to determine the grade.

The first question on the sheet is "Does this paper clearly (re)state the problem to be solved?" I search through the first few paragraphs. What catches my eye is the sentence, "In fact, I believe that Bobo was killed, and I will tell you how." This is a nice way of stating both the question and the answer in the same sentence. So I put a **YES** in the column for both questions number 1 and 2, and add a comment like

"last sentence of first paragraph. Good!"

Often students like to state the question at the beginning of the paper ("You want to know if Bobo killed himself or was murdered") and don't answer the question until the end of the paper. If that happens, I would still put a **YES** for the second question, but add a comment like,

"but not until the end of the paper. why not?"

If Todd had not included that last sentence of the first paragraph, I would have circled the "clearly" in the question. I might add a comment like, "first paragraph is vague about problem," and then put a **NO** in the column.

The next question is about physical assumptions. In this paper, reasonable physical considerations include things like air resistance, gusts of wind, and other material considerations that the mathematical model assumes, or fails to assume. Do I see anything in Todd's paper about air resistance? I skim through the next couple of paragraphs, but I don't see that phrase. (It also helps to look for the word "assume.") I also check the end of the paper for these words (students frequently put this near the end), but don't see it there. I'll have to read more carefully, but I add the comment,

"air resistance?"

under question three as a reminder to myself and to Todd. I don't write **NO** just yet; but I will later, after I read through more carefully and still don't find it.

I skip now to question 5 ("Does this paper clearly label diagrams … ?"). Todd's four figures have descriptive titles and ample labeling. I add the comment "nice!" and put a **YES** for this question. If there are no diagrams or tables, I'd put a slash or write **N.A.** meaning that this question will not count in the final tally.

Two other questions are fairly easy to answer after skimming the paper. One is number 8 ("Does this paper give acknowledgment where it is due?"). I look for references to a book or to people, often near the end of the paper. In Todd's paper, I see nothing. I write,

> *"Did you use no books? Talk to no one?"*

and put **NO** in the column. Sometimes students later come to explain that they did not, in fact, use any outside help, but I try to err on the pessimistic side. Are you sure you didn't ask anyone if they got the same answer as you? That's outside help. Are you sure you didn't use your book or your notes to come up with any of these formulas? That's outside help. Why didn't you get anyone to proofread your paper? I will change this **NO** later if I need to, though.

The other fairly easy question for me to answer by skimming is number 9 ("In this paper, are the spelling, grammar, and punctuation correct?"). Having skimmed the first couple of paragraphs, it seems that Todd's grammar and spelling seem fine so far. Otherwise I would have circled, say, "spelling," and added a comment:

> *"spell-check! always!"*

So here comes my own personal pet peeve: punctuation near equations. The place that students tend to mess up on punctuation is when they're doing long list of algebraic computations. Does Todd have any of those? Sure enough, I skim through and see, toward the end of the paper, two places where he lists three equations in a row, and those lists are not included in English sentences. (In his paper, they occur on pages 4 and 5.) I circle "punctuation" in the question, put a **NO** on this question, and write

> *"missing near equations, see below"*

in the comment section. At the bottom of the checklist, where there is more room for detailed comments, I write additional comments, indicating the question they refer to:

> *#9. See pages 4 and 5 in your paper. Equations need to be part of English sentences, with punctuation, words, etc. Read my Guide to Writing.*

The rest of the questions all need a somewhat more thorough reading. For question 4 ("Does this paper aim its explanations at the appropriate audience?"), Todd tries hard to make sure the police officer understands the math, but the paper is still overly algebraic. On the one hand, I like the description of the sockball to explain the lack of horizontal acceleration. Still, does Officer Kovalevskia understand phrases like "by integrating the second derivative . . . ?" Or does she want to see the step-by-step solving of a quadratic equation? I'd add comments to this effect, and give this question an **OK**, which counts as a **YES** in the grading, but serves as a warning that Todd could do better.

Todd doesn't pull any formulas out of a hat, or skip major steps; the mathematics is correct; and he answers the question that the Officer Kovalevskia set forth in her letter, so questions number 7, 10, and 11 all get a **YES** (as they usually do).

The one question left (number 6) is about defining variables. When Todd defines the variables for the initial velocity and acceleration, he's perhaps not as clear as I would like, but he does so using careful descriptions and also units of measurement. This is good. However, he defines "y'(0)" but not "y'(t)," and he does not define the variable "y" except with an oblique reference to vertical. (A clearer definition would be "The variable "y(t)" stands for Bobo's height above the cannon, measured in meters, t seconds after the cannon has been fired.") I write the comment

$y? t? y'(t)? x?$

and give question 6 a **NO**.

Now comes the time for tallying the grade. There are seven questions which got a **YES** or **OK**; I write "7" next to "grade" at the top of the sheet. All eleven questions counted this time (sometimes, some are **N.A.**—not applicable), so I write "11" next to "out of." That is Todd's grade: 7 out of 11, or 64%.

TABLE 1

Project grading checklist

---

grade ___7___

NAME: _____   out of ___11___

Checklist For Your Writing Project

Directions:

Please staple your paper and number the pages. Attach this page with a paper-clip to the front of your writing assignment when you turn it in. Your instructor will use this list to grade your assignment, and will return it to you with comments. Keep a copy of your paper for your own reference.

Please feel free to use this checklist as a guide for yourself while writing this assignment.

Does This Paper:

1. clearly (re)state the problem to be solved?   *YES*

2. state the answer in a complete sentence which stands on its own?   *YES*
   last sentence of first paragraph. Good!

3. clearly state the physical assumptions which underlie the formulas?   *NO*
   air resistance?

4. aim its explanations at the appropriate audience?   *OK*

5. clearly label diagrams, tables, graphs, or other visual representations of the math (if these are indeed used)?   *YES*
   nice!

6. define all variables used?   *NO*
   y? t? $y'(t)$? x?

7. explain how each formula is derived, or where it can be found?   *YES*

8. give acknowledgment where it is due?   *NO*
   Did you use no books? Talk to no one?

In this paper,

9. are the spelling, grammar, and punctuation correct?   *NO*
   missing near equations, see below

10. is the mathematics correct?   *YES*

11. did the writer solve the question that was originally asked?   *YES*

Comments:

#9. See pages 4 and 5 in your paper. Equations need to be part of English sentences, with punctuation, words, etc. Read my Guide to Writing.

3.2 Solution Paper and Rubric
—Gavin LaRose

The solution paper in this section is for the project "A Question of Law," one of Gavin LaRose's projects. This is a first project in a sophomore ordinary differential equations course, which requires that students develop and solve a first-order differential equation. Information about the distance an object is falling is given, and students must derive a relationship for $v(h)$, the velocity as a function of the height from which the body falls. Almost all students completing the project did so in the manner illustrated here, by finding an equation for $v'(t)$, solving this successively for $v(t)$ and $h(t)$, and then inverting to find $t(h)$ to allow the calculation of $v(h)$ $(= v(t(h)))$. Accordingly, the grading rubric that follows the solution paper is geared to those steps. One group determined (correctly) that there is a computationally easier solution method, admitted by rewriting the equation for $v'(t)$ as an equation for $v'(h)$ using the chain rule. The rubric provided would have to be modified for this case, and this is discussed briefly at the end of the section. Finally, it is worth noting that for this solution paper (or ones similar to it) the appendices were allowed to be neatly written by hand, rather than having the equations typeset using an equation editor.

In the subsections that follow are the sample solution paper, the rubric used to grade it, and comments regarding the grading.

3.2.1 Solution Paper

Rigorous Mathematical Contractors (RiMaC), Inc.
Suite 3, Strawmarket Business Plaza
Lonlinc, SK 04685
February 19xx

Hangemhi, Inc.
Suite 101, Boldledge Business Park
Lonlinc, SK 04685

Dear M. Arro:

You gave us information regarding a client of yours who may have fallen from a window and asked our company to find the velocity the client would have had upon striking the ground after falling 5, 15, 25 or 45 feet. You provided the data that the client is 180 lbs, and that terminal velocity for a falling body (person) is about 120 mph (176 ft/s). In this letter we describe our solution to your problem.

First, we will give an equation for the velocity of a falling object. Then, we solve this equation to find the velocity, and the time of fall of an object with respect to time. Using these formulas, we were able to find the time of fall of an object with respect to the height of the fall. Finally, we found the velocity of the body for the times of each of the heights of 5, 15, 25 and 45 feet.

Our physical measurements department has determined that the force on a falling body may be modeled by:

$$F = mg - kv^2$$

where F = force, v = velocity, m = mass, g = acceleration due to gravity, and K is some constant of air resistance. Because we know that $F = ma$, and that a, which is acceleration, is the derivative of velocity, we were able to rewrite our model as:

$$m\frac{dv}{dt} = mg - kv^2$$

Now with this and the data you gave us we were able to find K. Because acceleration equals zero at terminal velocity (v_T):

$$m \cdot 0 = mg - kv_T^2$$

$$K = \frac{mg}{v_T^2}$$

Plugging in $mg = 180$ lbs and $v_T = 176$ ft/s^2, $K = 0.005811$ lb $\cdot$ s^2/ft.

In rewriting our model, we were able to find a velocity function. First, we divided it by m. Next, we used separation of variables and rearranged the equation and integrated it to find a function of v with respect to t:

$$\int \frac{dv}{g - kv^2/m} = \int dt$$

We could solve this using partial fractions, see appendix 1. We found:

$$v(t) = \sqrt{mg/k} \ \ \frac{1 - Ce^{-2t\sqrt{gk/m}}}{1 + Ce^{-2t\sqrt{gk/m}}}$$

This is our equation for v as a function of time. However, you gave us the distance your client may have fallen, not how long he (or she) fell. So we found an equation for a value of time that depends on the height fallen.

We found the height fallen with respect to time, $h(t)$, because we know the integral of $v(t)$ is $h(t)$. After simplifying $v(t)$, see appendix 3, we used a program called *Mathematica* to integrate $v(t)$ and find $h(t)$. We got:

$$h(t) = \int v(t)\, dt = \int v_T \frac{1 - Ce^{-2t\sqrt{gk/m}}}{1 + Ce^{-2t\sqrt{gk/m}}}$$

$$h(t) = \frac{v_T^2}{g} \ln\left(e^{2gt/v_T} + 1\right) - v_T t + C$$

However, we had to get rid of the constant of integration, C. To do this, we used the initial condition $h(0) = 0$. For our model downwards is the positive direction, therefore $h(0) = 0$, see appendix 4.

Now we want to know the time fallen depending on the height, 5, 15, 25 or 45 feet. We solved for t in our equation for $h(t)$, see appendix 5, to obtain an equation for $t(h)$:

$$t(h) = \frac{v_T}{g} \quad \mathrm{arccosh}(e^{gh/v_T^2})$$

Using the values you gave us and $g = 32$ ft/s² we were able to answer your question. We found the times for each height with $t(h)$ and then plugged into $v(t)$ to get the values in the table below.

Height of fall: (h)	Time of fall ($t(h)$)	Final velocity: ($v(t(h))$)
5 feet	0.55950 sec	17.8425 feet/sec
15 feet	0.97075 sec	30.7454 feet/sec
25 feet	1.25539 sec	39.4890 feet/sec
45 feet	1.69008 sec	52.4421 feet/sec

These are the velocities your client probably would have reached in the course of his or her precipitous descent from the window(s) you asked about. We hope this information is useful to you, and hope that if you have mathematical needs in the future you will contact our company again.

Yours sincerely,
Math Students
Rigorous Mathematical Contractors (RiMaC), Inc.

Appendices

Partial Fractions

$$\int \frac{dv}{g - \frac{k}{m}v^2} = \int \frac{dv}{(\sqrt{g} - \sqrt{k/m}v)(\sqrt{g} + \sqrt{k/m}v)} = \int dt$$

With partial fractions we split this term up:

$$\int \frac{A\,dv}{\sqrt{g} - \sqrt{k/m}v} + \int \frac{B\,dv}{\sqrt{g} + \sqrt{k/m}v} = \int dt$$

$$-\sqrt{\frac{m}{k}} A \ln\left(\sqrt{g} - \sqrt{\frac{m}{k}}v\right) + \sqrt{\frac{m}{k}} A \ln\left(\sqrt{g} + \sqrt{\frac{m}{k}}v\right) = t + C$$

Solving for A and B to make the fraction and split fraction the same:

$$A(\sqrt{g} + \sqrt{k/mv}) + B(\sqrt{g} - \sqrt{k/mv}) = 1$$

(by finding a common denominator).

$$(A + B)\sqrt{g} + (A - B)\sqrt{\frac{k}{m}}v = 1,$$

so

$$(A + B) = \frac{1}{\sqrt{g}}, \quad \text{and} \quad A - B = 0 \quad \text{or} \quad A = B$$

Substituting A for B:

$$2A = \frac{1}{\sqrt{g}} \quad A = \frac{1}{2\sqrt{g}} = B.$$

And plugging these into the equation:

$$-\frac{1}{2\sqrt{g}}\sqrt{\frac{m}{k}} \ln\left(\sqrt{g} - \sqrt{\frac{k}{m}}\right) + \frac{1}{2\sqrt{g}}\sqrt{\frac{m}{k}} \ln\left(\sqrt{g} + \sqrt{\frac{k}{m}}\right) = t + C$$

$$\ln\left(\sqrt{g} - \sqrt{\frac{k}{m}}\right) - \ln\left(\sqrt{g} + \sqrt{\frac{k}{m}}\right) = -2\sqrt{\frac{gk}{m}}\,t + C$$

$$\exp\left(\ln\left(\sqrt{g} - \sqrt{\frac{k}{m}}\right) - \ln\left(\sqrt{g} + \sqrt{\frac{k}{m}}\right)\right) = \exp\left(-2\sqrt{\frac{gk}{m}}\,t + C\right)$$

$$\frac{\sqrt{g} - \sqrt{\frac{k}{m}}v}{\sqrt{g} + \sqrt{\frac{k}{m}}v} = Ce^{-2t\sqrt{gk/m}},$$

$$\sqrt{g} - \sqrt{\frac{k}{m}}v = \left(\sqrt{v} + \sqrt{\frac{k}{m}}v\right)(Ce^{-2t\sqrt{gk/m}})$$

$$\sqrt{g} - \sqrt{\frac{k}{m}}v = C\sqrt{g}\,e^{-2t\sqrt{gk/m}} + C\sqrt{\frac{k}{m}}\,e^{-2t\sqrt{gk/m}}$$

$$v\sqrt{\frac{k}{m}}\left(Ce^{-2t\sqrt{gk/m}} + 1\right) = \sqrt{g}\left(1 - Ce^{-2t\sqrt{gk/m}}\right)$$

$$v(t) = \sqrt{\frac{mg}{k}} \cdot \frac{1 - Ce^{-2t\sqrt{gk/m}}}{1 + Ce^{-2t\sqrt{gk/m}}}$$

Solving for C in $v(t)$ Our initial condition is $v(0) = 0$. Therefore,

$$v(0) = 0 = \sqrt{\frac{mg}{k}} \cdot \frac{1 - Ce^{-2\cdot 0\sqrt{gk/m}}}{1 + Ce^{-2\cdot 0\sqrt{gk/m}}}$$

$$0 = \frac{1 - C}{1 + C}$$

$$C = 1$$

Therefore,

$$v(t) = \sqrt{\frac{mg}{k}} \cdot \frac{1 - e^{-2t\sqrt{gk/m}}}{1 + e^{-2t\sqrt{gk/m}}} \cdot$$

Simplifying $v(t)$

$$v(t) = v_T \cdot \frac{1 - e^{-2t\sqrt{g/v_T}}}{1 + e^{-2t\sqrt{g/v_T}}}$$

$$v(t) = v_T \cdot \frac{e^{2t\sqrt{g/v_T}} - 1}{e^{2t\sqrt{g/v_T}} + 1}$$

Solving for C in $h(t)$

$$h(t) = \frac{v_T^2}{g} \cdot \ln\left(e^{2gt/v_T} + 1\right) - v_T t + C$$

since down is positive, $h(0) = 0$.

$$0 = \frac{v_T^2}{g} \cdot \ln(2) + C$$

$$C = -\frac{v_T^2}{g} \ln(2)$$

so

$$h(t) = \frac{v_T^2}{g} \cdot \ln\left(e^{2gt/v_T} + 1\right) - v_T t - \frac{v_T^2}{g} \ln(2)$$

Solving for $t(h)$ from $h(t)$

$$h = \frac{v_T^2}{g} \cdot \ln\left(e^{2gt/v_T} + 1\right) - v_T t - \frac{v_T^2}{g} \ln(2)$$

$$\frac{gh}{v_T^2} = \ln\left(e^{2gt/v_T} + 1\right) - \frac{gt}{v_T} - \ln(2)$$

$$e^{gh/v_T^2} = \left(e^{2gt/v_T} + 1\right) \cdot e^{-gt/v_T} \cdot e^{-\ln 2}$$

$$2e^{gh/v_T^2} = e^{gt/v_T} + e^{-gt/v_T}$$

$$2e^{gh/v_T^2} = 2\cosh\left(\frac{gt}{v_T}\right)$$

$$e^{gh/v_T^2} = \cosh\left(\frac{gt}{v_T}\right)$$

$$\frac{gt}{v_T} = \operatorname{arccosh}(e^{gh/v_T^2})$$

$$t = \frac{v_T}{g}\operatorname{arccosh}(e^{gh/v_T^2})$$

3.2.2 Rubric

As described in the introduction, the rubric to grade a project is determined by isolating the steps it requires and assigning point values to these. The rubric following in Table 3 does this, according to the steps outlined in the preceding sections for this project. In each case, we determine the point values for each objective in the rubric according to an analysis of what the correct solution requires. More explanation of rubrics and their development is found in Section 1.4.2.

3.2.3 Grading and Comments

There are a couple of problems with this solution paper. The authors make some errors in verbalizing in mathematical terms what they are finding (in particular in the introductory paragraph), and the paper has some rough corners in its clarity of presentation and correct use of mathematical prose. Appendix 2 is never referred to, and the material in appendix 3 does not appear to do what it purports to do (simplify the expression for $v(t)$ to make it easier to integrate to find $h(t)$). In the calculation for $v(t)$ we are not told that an initial condition was applied to find the constant of integration C (similarly, while there is mention of the initial condition for $h(t)$ it is not well explained). However, the mathematical calculations themselves are (with those exceptions) correct—in much the manner we frequently find that those of science and engineering students are. Finally, the development of the final result, the actual velocities attained by the falling client, is rather terse. However, taken as a whole, the paper is quite respectable.

Using the rubric in Table 2, we might therefore give the students 2 points for their model development, 3 for the determination of K, 3 for the solution for $v(t)$, 3 for finding $h(t)$, 2 for for $v(h)$, and 2 for clarity and 2 for meeting deadlines (assuming that they were met), for a total of 17/20. If it were a later project, the ambiguity in the solution for $h(t)$ might also draw a one point reduction. In either case, this is a very reasonable "B" type solution—with some work and cleaning up, it would easily become an excellent paper. Feedback given to the students would be the rubric with their scores on it, and their solution paper with comments indicating the criticisms above (as well as corrections to such grammatical errors as seemed most egregious on reading it) marked.

Finally, we noted earlier that the problem can be solved differently, by rewriting the equation for v with independent variable h (by noting that $dv/dt = (dv/dh) \cdot (dh/dt) =$

$v'(h) \cdot v$). In this case, the middle two rows in the rubric (finding $v(t)$ and $h(t)$) aren't applicable. However, the calculation of $v(h)$ is much more important. Therefore, we might assign up to four points for the calculation of $v(h)$ (as with $v(t)$ in the indicated rubric). In addition, we need to assess students' application of the chain rule to rewrite the model, leading to an objective "Rewriting Model" (for, say, up to three points). Then, because in this case the calculation of $v(h)$ is distinct from the application of the given data (the use of the data is implied in our original rubric), we might also add the objective "Find Specific $v(h)$," and give up to two points for stepping correctly between the general solution for $v(h)$ and $v(h)$ for $h = 5, 15, 25$, and 45 feet. Finally, because it is more elegant than the other solution method, we might award the last point for the students' having correctly determined this way of solving the problem.

TABLE 2
Project grading rubric

Objective	0 points	1 point	2 points	3 points	4 points
Model Development	no sensible model	reference to the model, incorrect incorporation of dv/dt, and/or poor or missing explanation.	correct model, well explained		
Solution for K	no correct calculation for K	reference to the model and v_T or a, but incorrect inclusion of one or both; poor explanation	use of model, inclusion of v_T and a, minor errors in calculation or clarity	correct solution for K	
Solution for $v(t)$	no correct solution for $v(t)$ or no explanation	mention of separation of variables but no or incorrect solution	use and explanation of separation of variables, but gaps in explanation, or significant algebraic or integration errors	use and explanation of separation of variables with only small gaps or minor errors	correct, well explained solution
Solution for $h(t)$	no solution, or no explanation	link made between $h(t)$ and $v(t)$, but no solution, or some solution with major errors or incomplete sections	link made between $h(t)$ and $v(t)$, some calculations, but with errors in integration or unclear explanation.	correct, well explained solution	
Calculation of $v(h)$	no calculation, or no explanation	$t(h)$ found, but use to find $v(h)$ incorrect or unclear, or vice-versa, with poor explanation	one of $t(h)$ or $v(h)$ found correctly, but small errors in the other, or missing explanation	correct solutions for $t(h)$ and $v(h)$	
Clarity and Organization	utterly unclear or completely disorganized paper	multiple unclear sections or limited organization	small unclear section(s) or spotty organization	crystal clear paper	
Deadlines	all deadlines missed	one or more deadlines missed	all deadlines met		

4

Additional Sample Checklists and Rubrics

TABLE 1
Tommy Ratliff's checklist

NAMES: _____ GRADE: _____

_____ OUT OF: 100

Directions:

- Please attach this page with a paper-clip to your writing assignment when you turn it in.
- This list will be used to grade your assignment, and will be returned to you with comments.
- Please feel free to use this checklist as a guide for yourself while writing the assignment.

Does this paper:

1. clearly (re)state the problem to be solved?
2. provide a paragraph which explains how the problem will be approached?
3. state the answer in a few complete sentences which stand on their own?
4. give a precise and well-organized explanation of how the answer was found?
5. clearly label diagrams, tables, graphs, or other visual representations of the math?
6. define all variables, terminology, and notation used?
7. clearly state the assumptions which underlie the formulas and theorems, and explain how each formula or theorem is derived, or where it can be found?
8. give acknowledgment where it is due?
9. use correct spelling, grammar, and punctuation?
10. contain correct mathematics?
11. solve the questions that were originally asked?

Comments:

TABLE 2
Elyn Rykken's checklist

Instructions given to students with the checklist:

- Please attach this page with a paper-clip to your writing assignment when you turn it in.
- This list will be used to grade your assignment and will be returned to you with comments. Keep a copy of your paper for your own reference
- Use this checklist as a guide for yourself while writing the assignment

Form:	5 points
Does this paper:	1. clearly (re)state the problem to be solved (including on the essential details)?
	2. explain what level and types of math will be used?
	3. solve the question that was originally asked? (2 pts)
	4. give acknowledgment where it is due (did anyone work with you on the math)?
Content:	**7 points**
Does this paper:	5. give a precise and well-organized explanation of how the answer was found? (2 pts)
	6. define all variables, terms, and notation used?
	7. explain how each formula is derived or where it can be found?
	8. clearly label diagrams, tables, graphs or other visual representations of the math?
	9. contain correct mathematics? (2 pts)
Presentation:	**3 points**
Does this paper:	10. use standard business letter form?
	11. use correct spelling, grammar, and punctuation?
	12. look neat? (typing helps with this)
Comments:	

TABLE 3

Gavin LaRose's rubric for "A Jump and a Jerk"

Objective	0 points	1 point	2 points	3 points
Model Development	no sensible model	one term (mv' or a force) in model correct	two terms correct, or three but with errors in finding parameters	correct model
Deployment Time	no correct model solution	solution partially relating velocity to distance fallen and deployment time	solution relating velocity to distance fallen and deployment time, with numerical or minor conceptual errors	correct calculation
Landing Location	no sensible model or solution	partially correct model, solution for distance traveled with some conceptual or numerical error	correct model ($x'' =$ drag force), with solution error, or error in model but correct solution for distance traveled	correct model and solution
Smoothing Function	no smoothing function	correct polynomial form for smoothing function or correct conditions to determine constants	correct form for function, correct conditions, error(s) in calculations	correct calculation
Jerk Calculation	no intelligible calculation	conceptual errors in jerk calculation	correct conceptual calculation, numerical errors	correct calculation
Clarity and Organization	utterly unclear or completely disorganized paper	multiple unclear sections or limited organization	small unclear section(s) or spotty organization	crystal clear paper
Deadlines	all deadlines missed	one or more deadlines missed	all deadlines met	N/A

5

Index by Course

Multivariable Calculus

Differential Equations

Author Biographies

Annalisa Crannell is an Associate Professor and Chair of Mathematics at Franklin & Marshall College. She teaches courses at all levels of the undergraduate mathematics curriculum, and she does research in the area of topological dynamical systems. She has worked on numerous committees of the AMS and MAA, and in particular is an editor of the Mathematical Association of America's "Notes" series of books, and an associate editor of Mathematics Magazine. In addition to her purely mathematical research, Annalisa Crannell is well known for her work in writing across the curriculum and for her work in mathematics and art. Her classroom materials have been adapted and adopted by more than 100 high school and college instructors across the nation.

Gavin LaRose received his PhD in applied math (officially, "engineering sciences and applied math") at Northwestern University. He has taught in a small liberal arts college environment and now works in instructional technology and teaches at the University of Michigan. His use of projects spans eight years and runs the gamut from calculus I to linear algebra and differential equations. He has won teaching awards as a graduate student and faculty member, and is an associate co-director of the MAA's Project NExT.

Thomas (Tommy) Ratliff is an Associate Professor at Wheaton College in Norton, Massachusetts. He did his graduate work at Northwestern University in algebraic topology, and his current area of research is voting theory. Tommy taught at Kenyon College and St. Olaf College before he started at Wheaton in 1996. He is interested in using writing projects in (almost) all of his classes, and he has given numerous presentations at regional and national meetings on pedagogical topics. Tommy participated in the first group of Project NExT Fellows in 1994–1995.

Elyn Rykken is an Assistant Professor at Muhlenberg College in Allentown, Pennsylvania. She received her PhD from Northwestern University, studying dynamical systems. Before starting at Muhlenberg College in 1999, Elyn taught at Indiana University Northwest. She has used projects in a survey of mathematics course and currently uses them in her calculus classes. Elyn was a 1995–1996 Project NExT Fellow and a 1997 participant of the Institute in the History of Mathematics and Its Use in Teaching.